Research in Metabolomics via Nuclear Magnetic Resonance Spectroscopy: Data Mining, Biochemistry and Clinical Chemistry

Research in Metabolomics via Nuclear Magnetic Resonance Spectroscopy: Data Mining, Biochemistry and Clinical Chemistry

Editors

Alessia Vignoli
Gaia Meoni
Leonardo Tenori

MDPI • Basel • Beijing • Wuhan • Barcelona • Belgrade • Manchester • Tokyo • Cluj • Tianjin

Editors
Alessia Vignoli
University of Florence
Italy
Consorzio Interuniversitario
Risonanze Magnetiche
MetalloProteine (CIRMMP)
Italy

Gaia Meoni
University of Florence
Italy
Consorzio Interuniversitario
Risonanze Magnetiche
MetalloProteine (CIRMMP)
Italy

Leonardo Tenori
University of Florence
Italy
Consorzio Interuniversitario
Risonanze Magnetiche
MetalloProteine (CIRMMP)
Italy

Editorial Office
MDPI
St. Alban-Anlage 66
4052 Basel, Switzerland

This is a reprint of articles from the Special Issue published online in the open access journal *Applied Sciences* (ISSN 2076-3417) (available at: https://www.mdpi.com/journal/applsci/special_issues/metabolomics_via_nuclear_magnetic_resonance_spectroscopy).

For citation purposes, cite each article independently as indicated on the article page online and as indicated below:

LastName, A.A.; LastName, B.B.; LastName, C.C. Article Title. *Journal Name* **Year**, *Volume Number*, Page Range.

ISBN 978-3-0365-4553-0 (Hbk)
ISBN 978-3-0365-4554-7 (PDF)

Contents

About the Editors

Alessia Vignoli

Alessia Vignoli (Researcher), born in 1988 in Florence, earned her M.Sc. in Chemical Sciences in 2014 at the University of Florence. In 2017, she obtained her Ph.D. in Structural Biology, which is an international doctorate awarded by the University of Florence in collaboration with the Universities of Frankfurt and Utrecht, defending her thesis entitled "Applications of metabolomics in biomedicine". In 2018, she was the recipient of a two-year fellowship of the Italian Association for Cancer Research (AIRC) with a project focused on the applications of NMR-based metabolomics for the prognosis of colon cancer patients. From 2021, she has been a postdoctoral research associate at Department of Chemistry "Ugo Schiff", University of Florence, working on the risk stratification of atrial fibrillation patients using NMR-based metabolomics. Alessia Vignoli is author of 31 scientific articles, all related to metabolomics, and all published in international peer-reviewed journals.

Gaia Meoni

Gaia Meoni (Researcher) was born in 1991 in Florence. She obtained the M.Sc. in Molecular Biology and Genetics at the University of Pavia in 2015. In 2018, she earned the International Doctorate in Structural Biology from the University of Florence, in conjunction with the Universities of Frankfurt and Utrecht, defending a thesis titled "Metabolomics by NMR: applications and challenges from biomedicine to food research". After the Ph.D., she continued doing research in the field of metabolomics during her post-doctorate fellowships at CIRMMP (Interuniversity Consortium for Magnetic Resonance of Metallo Proteins) in 2019 and GiottoBiotech s.r.l from 2019 to 2021. Currently, she is junior researcher at the Department of Chemistry "Ugo Schiff" of the University of Florence. Her research activity focuses on NMR based metabolomics applications, which range from human health to food matrices. She is co-author of 23 peer-reviewed research articles in the field of metabolomics.

Leonardo Tenori

Leonardo Tenori (Assistant Professor), born in 1977, obtained his master's degree in chemistry in 2002 at the University of Florence, and the International Ph.D. in Structural Biology in 2008 at the University of Florence. He has been Researcher at CIRMMP (University of Florence, Italy), and he joined the faculty at the Department of Chemistry of the University of Florence as tenure-track researcher in September 2020. After the Ph.D., his primary interest turned to the applications of metabolomics via magnetic resonance spectroscopy (NMR) for the study of complex biological mixtures, both in the biomedicine field (analysis of human biofluids), and in the agri-food area. He performs his research in the framework of the CERM/CIRMMP infrastructure, one of the largest and best equipped NMR facilities in world. In 2015 he was awarded with a Fellowship of the Italian Foundation "Veronesi" for the study of the metabolomic alterations in melanoma patients. He has collaboration with researchers and clinicians at local, national and international level. He is co-author of 117 scientifical articles related to metabolomics in peer-review journals.

Preface to "Research in Metabolomics via Nuclear Magnetic Resonance Spectroscopy: Data Mining, Biochemistry and Clinical Chemistry"

It is our pleasure to present this Special Issue entitled "Research in Metabolomics via Nuclear Magnetic Resonance Spectroscopy: Data Mining, Biochemistry and Clinical Chemistry", which broadly addresses the applications of nuclear magnetic resonance (NMR) spectroscopy in the metabolomics field.

Metabolomics is defined as the comprehensive characterization of the ensemble of endogenous and exogenous metabolites present in a biological specimen. Metabolites represent, at the same time, the downstream output of the genome and the upstream input from various exogenous factors, such as the environment, lifestyle, and diet. Even though the first scientific paper explicitly dealing with metabolomics is more than 20 years old, we think that this collection is still timely and of particular interest for the scientific community because this "-omic" science is still growing and novel practical applications in biomedicine and in the agricultural field continue to emerge. As researchers at the Magnetic Resonance Center of the University of Florence (Italy) we decided to focus this Special Issue on NMR-based metabolomics, which is our main field of research. The ensemble of the studies present in this volume offers a representative overview of the applications of NMR metabolomics raging from the biomedical fields to food science.

In the end, we want to thank the authors for their precious contributions as this Special Issue would not be possible without them. We would also like to express our sincere appreciation to the dedicated editorial team of Applied Sciences for their valuable contributions to this volume.

Alessia Vignoli, Gaia Meoni, and Leonardo Tenori
Editors

Editorial

Applications and Challenges for Metabolomics via Nuclear Magnetic Resonance Spectroscopy

Alessia Vignoli [1,2,3,*], Gaia Meoni [1,2,3,*] and Leonardo Tenori [1,2,3,*]

1 Magnetic Resonance Center (CERM), University of Florence, 50019 Sesto Fiorentino, Italy
2 Department of Chemistry "Ugo Schiff", University of Florence, 50019 Sesto Fiorentino, Italy
3 Consorzio Interuniversitario Risonanze Magnetiche MetalloProteine (CIRMMP), 50019 Sesto Fiorentino, Italy
* Correspondence: vignoli@cerm.unifi.it (A.V.); meoni@cerm.unifi.it (G.M.); tenori@cerm.unifi.it (L.T.)

Citation: Vignoli, A.; Meoni, G.; Tenori, L. Applications and Challenges for Metabolomics via Nuclear Magnetic Resonance Spectroscopy. *Appl. Sci.* **2022**, *12*, 4655. https://doi.org/10.3390/app12094655

Received: 29 April 2022
Accepted: 4 May 2022
Published: 6 May 2022

1. Introduction

Even though metabolomics is about 20 years old, the interest in this "-omic" science is still growing, and high expectations remain in the scientific community for new practical applications in biomedicine and in the agricultural field. Thus far, biomedical metabolomic studies have produced great advancements in biomarker discovery, identification of novel metabolites, and more detailed characterization of biological pathways involved in the manifestation and progression of diseases. In parallel, metabolomics has been shown to have an emerging role in monitoring the influence of different manufacturing procedures on food quality and food safety. In light of the above, this Special Issue was introduced to collect the latest research from various application fields of NMR-based metabolomics [1,2], ranging from biomedicine to data mining and food chemistry.

2. NMR-Based Metabolomics

Our collection comprises four research articles that report interesting applications of NMR metabolomics in the biomedical setting. In the first article published in our issue, Baranovicova et al. [3] present a longitudinal study that explores the dynamics of metabolomic changes in the plasma of 53 patients, diagnosed with SARS-CoV-2 infection, at three consecutive time points during their first week of hospitalization (days 1, 3, and 7 after admission to the hospital) to reveal the differences among patients with positive (survivors) and negative (worsening condition, non-survivors) outcomes. People with COVID-19, regardless their prognosis, presented alterations in their energy and amino acids metabolism. These changes were normalized by the seventh day in patients with positive outcomes; conversely, they were not reverted in patients with negative outcomes. These results indicate that the ability to respond to metabolomic alterations induced by severe inflammation due to SARS-CoV-2 infection is a key factor in determining patients' outcomes and that these metabolic changes can be tackled with individual pharmacological or diet interventions to support patient response.

In recent years, nanoscience and nanotechnology have been developing rapidly; at the same time, the increased use of nanoparticles has raised several concerns regarding human public health and occupational safety. In the article by Horník et al. [4], NMR-based metabolomics of exhaled breath condensate (EBC) and blood plasma is used to study the effects of occupational exposure to nanoparticles (NPs). The EBC and blood plasma samples from 20 workers exposed to NPs were collected pre-shift (i.e., before 2.5 h of exposure to NPs) and post-shift (i.e., after NP exposure). Moreover, 20 controls (not exposed to NPs) were enrolled for this study. Multivariate statistical analyses, performed both on EBC and plasma NMR data, showed clear discriminations among the three groups of interest (the pre-shift, post-shift, and control groups). The univariate metabolite analysis revealed several alterations in subjects exposed to NPs, in particular the acute effect of NP exposure is primarily reflected in the metabolic pathways involved in the production of antioxidants and of other protective species, whereas the chronic effect of NP exposure seems to be associated with alterations in glutamine and glutamate metabolism, and the purine metabolism pathways.

The paper authored by Vignoli et al. [5] characterizes the effects of surgery on the serum metabolomic profiles of colorectal cancer (CRC) patients and explores the possibility that metabolic variations among preoperative and postoperative serum samples could be informative on future cancer recurrence. A total of 41 patients diagnosed with early-stage CRC and scheduled for radical resection were enrolled for this study. Serum samples collected preoperatively (t0) and 4–6 weeks after surgery but before the start of any treatment (t1) were analyzed via ^{1}H NMR spectroscopy. A clear discrimination between t0 and t1 emerged: after surgery, there are significant increases in pyruvate, HDL cholesterol, HDL phospholipids, HDL Apo-A1, and HDL Apo-A2 levels, coupled with significant decreases in acetone, 3-hydroxybutyrate, LDL-Chol/HDL-Chol ratio, and Apo-A1/Apo-B100 ratio. Taken together, these results point to a relevant rearrangement of the metabolic pathways related to lipoproteins, ketone bodies, and energy metabolism. Furthermore, several differences between post- and pre-operative serum samples, in particular those related to the HDL-Chol and VLDL-Chol subfractions, seem to be associated with cancer recurrence. These data pave the way for novel strategies for risk stratification in patients with early-stage CRC.

The paper by Georgiopoulou et al. [6] is the last research article related to biomedical applications of NMR metabolomics included in our issue. It proposes an analysis of urine samples of preterm infants with neonatal sepsis, a systemic infection difficult to diagnose in its early stages and thus reporting high rates of morbidity and mortality. In this study, the urine metabolomic profiles of 34 septic neonates, 14 preterm neonates without sepsis or other serious morbidity but hospitalized in the NICU, and 23 healthy preterm neonates were examined. Multivariate and univariate statistical analyses showed clear discriminations between septic and healthy newborns. In particular, alterations in the levels of gluconate, myo-inositol, hippurate, taurine, N, N-Dimethylglycine, betaine, creatinine, glucose and lactose emerged as the most significant. These data represent a promising basis for future large-scale multicenter studies and give new perspectives for clinical research in the field of neonatology.

We decided to address also foodomics in our issue, which refers to metabolomic approaches applied to foodstuff for investigating topics related to nutrition, fraud detection and traceability of the geographical origin and production/processing procedures of food. In this regard, in our issue, we decided to publish an NMR-based metabolomic study based on water extracts of green and roasted coffee beans of different cultivars from three distinct Nicaraguan farms [7]. We think that this study can show well the potential and versatility of NMR metabolomics. Here, the authors demonstrate the potential of NMR metabolomics not only to define the geographical origin and the farm of provenance but also to characterize the effect of the environment (microclimates, irrigation, fertilizers, etc.) and the post-harvest practices (e.g., drying and fermentation) that are responsible for different aroma precursors in coffee and thus affect its distinct taste.

The ensemble of these studies offers a representative overview of the applications of NMR metabolomics raging from the biomedical fields to food science.

The capabilities of NMR, coupled with an ever-growing list of statistical chemometric techniques, make NMR-based metabolomics a versatile technique. Applying correct and suitable statistical techniques has become of fundamental importance for metabolomics studies. For this reason the review of Corsaro et al. [8], which lists some of the most commonly used and useful statistical techniques in metabolomics, explaining their advantages and disadvantages, has been included in our issue. In this work, the authors give an overview of the wide range of statistical opportunities for NMR-based metabolomics, ranging from conventional approaches (e.g., unsupervised and supervised methods, and pathway analyses) to less frequently applied deep learning and artificial neural networks. We found this review beneficial not only for fledgling metabolomic students approaching chemometrics but also for experts in the field looking for a more suitable approach to their studies.

In conclusion, the current Special Issue of *Applied Sciences* offers a variety of examples on how NMR-based metabolomics can potentially be used in several and varied settings. Although this Special Issue has been closed, more in-depth research on this topic is expected in the years to come, and future research will no doubt continue to explore the possibility of translating metabolomics into real-life applications.

Conflicts of Interest: The authors declare no conflict of interest.

References

1. Vignoli, A.; Ghini, V.; Meoni, G.; Licari, C.; Takis, P.G.; Tenori, L.; Turano, P.; Luchinat, C. High-Throughput Metabolomics by 1D NMR. *Angew. Chem. Int. Ed. Engl.* **2019**, *58*, 968–994. [CrossRef] [PubMed]
2. Takis, P.G.; Ghini, V.; Tenori, L.; Turano, P.; Luchinat, C. Uniqueness of the NMR Approach to Metabolomics. *TrAC Trends Anal. Chem.* **2019**, *120*, 115300. [CrossRef]
3. Baranovicova, E.; Bobcakova, A.; Vysehradsky, R.; Dankova, Z.; Halasova, E.; Nosal, V.; Lehotsky, J. The Ability to Normalise Energy Metabolism in Advanced COVID-19 Disease Seems to Be One of the Key Factors Determining the Disease Progression—A Metabolomic NMR Study on Blood Plasma. *Appl. Sci.* **2021**, *11*, 4231. [CrossRef]
4. Horník, Š.; Michálková, L.; Sýkora, J.; Ždímal, V.; Vlčková, Š.; Dvořáčková, Š.; Pelclová, D. Effects of Workers Exposure to Nanoparticles Studied by NMR Metabolomics. *Appl. Sci.* **2021**, *11*, 6601. [CrossRef]
5. Vignoli, A.; Mori, E.; Di Donato, S.; Malorni, L.; Biagioni, C.; Benelli, M.; Calamai, V.; Cantafio, S.; Parnofiello, A.; Baraghini, M.; et al. Exploring Serum NMR-Based Metabolomic Fingerprint of Colorectal Cancer Patients: Effects of Surgery and Possible Associations with Cancer Relapse. *Appl. Sci.* **2021**, *11*, 11120. [CrossRef]
6. Georgiopoulou, P.D.; Chasapi, S.A.; Christopoulou, I.; Varvarigou, A.; Spyroulias, G.A. Untargeted 1H-NMR Urine Metabolomic Analysis of Preterm Infants with Neonatal Sepsis. *Appl. Sci.* **2022**, *12*, 1932. [CrossRef]
7. Meoni, G.; Luchinat, C.; Gotti, E.; Cadena, A.; Tenori, L. Phenotyping Green and Roasted Beans of Nicaraguan Coffea Arabica Varieties Processed with Different Post-Harvest Practices. *Appl. Sci.* **2021**, *11*, 11779. [CrossRef]
8. Corsaro, C.; Vasi, S.; Neri, F.; Mezzasalma, A.M.; Neri, G.; Fazio, E. NMR in Metabolomics: From Conventional Statistics to Machine Learning and Neural Network Approaches. *Appl. Sci.* **2022**, *12*, 2824. [CrossRef]

Article

The Ability to Normalise Energy Metabolism in Advanced COVID-19 Disease Seems to Be One of the Key Factors Determining the Disease Progression—A Metabolomic NMR Study on Blood Plasma

Eva Baranovicova [1], Anna Bobcakova [2], Robert Vysehradsky [2], Zuzana Dankova [1], Erika Halasova [1], Vladimir Nosal [3] and Jan Lehotsky [4,*]

[1] Biomedical Center Martin, Jessenius Faculty of Medicine in Martin, Comenius University in Bratislava, Mala Hora 4, 036 01 Martin, Slovakia; eva.baranovicova@uniba.sk (E.B.); zuzana.dankova@uniba.sk (Z.D.); erika.halasova@uniba.sk (E.H.)
[2] Clinic of Pneumology and Phthisiology, Jessenius Faculty of Medicine and University Hospital in Martin, Comenius University in Bratislava, Mala Hora 4, 036 01 Martin, Slovakia; anna.bobcakova@uniba.sk (A.B.); robert.vysehradsky@uniba.sk (R.V.)
[3] Clinic of Neurology, Jessenius Faculty of Medicine and University Hospital in Martin, Comenius University in Bratislava, Mala Hora 4, 036 01 Martin, Slovakia; vladimir.nosal@uniba.sk
[4] Department of Medical Biochemistry, Jessenius Faculty of Medicine in Martin, Comenius University in Bratislava, Mala Hora 4, 036 01 Martin, Slovakia
* Correspondence: jan.lehotsky@uniba.sk

Citation: Baranovicova, E.; Bobcakova, A.; Vysehradsky, R.; Dankova, Z.; Halasova, E.; Nosal, V.; Lehotsky, J. The Ability to Normalise Energy Metabolism in Advanced COVID-19 Disease Seems to Be One of the Key Factors Determining the Disease Progression—A Metabolomic NMR Study on Blood Plasma. *Appl. Sci.* **2021**, *11*, 4231. https://doi.org/10.3390/app11094231

Academic Editor: Alessia Vignoli

Received: 8 April 2021
Accepted: 4 May 2021
Published: 7 May 2021

Abstract: Background: COVID-19 represents a severe inflammatory condition. Our work was designed to monitor the longitudinal dynamics of the metabolomic response of blood plasma and to reveal presumable discrimination in patients with positive and negative outcomes of COVID-19 respiratory symptoms. Methods: Blood plasma from patients, divided into subgroups with positive (survivors) and negative (worsening condition, non-survivors) outcomes, on Days 1, 3, and 7 after admission to hospital, was measured by NMR spectroscopy. Results: We observed changes in energy metabolism in both groups of COVID-19 patients; initial hyperglycaemia, indicating lowered glucose utilisation, was balanced with increased production of 3-hydroxybutyrate as an alternative energy source and accompanied by accelerated protein catabolism manifested by an increase in BCAA levels. These changes were normalised in patients with positive outcome by the seventh day, but still persisted one week after hospitalisation in patients with negative outcome. The initially decreased glutamine plasma level normalised faster in patients with positive outcome. Patients with negative outcome showed a more pronounced Phe/Tyr ratio, which is related to exacerbated and generalised inflammatory processes. Almost ideal discrimination from controls was proved. Conclusions: Distinct metabolomic responses to severe inflammation initiated by SARS-CoV-2 infection may serve towards complementary personalised pharmacological and nutritional support to improve patient outcomes.

Keywords: NMR metabolomics; human plasma; COVID-19

1. Introduction

COVID-19, which develops after SARS-CoV-2 infection, represents a severe inflammatory condition. Over the past two decades, a close link between metabolism and immunity has emerged [1,2]. The immune reaction in severe inflammation is intimately associated with a dependency on amino acids included in the proteosynthesis and specific metabolism of immunocompetent cells [3]. In addition, the immune response of the organism is also closely related to glucose energetical metabolism [1,2,4–6]. Synergic interactions between metabolism and immune processes serve as a tool to monitor the particular state of an organism relating to immunological response via metabolomics analysis. The increasing number of studies confirms the great potential of the metabolomic approach in the evaluation of COVID-19 disease, its course, and its outcome [7–10]. A comprehensive metanalysis of COVID-19 patients showed several key metabolic characteristics for disease progression and clinical outcome [11]. Untargeted metabolomics on patients' serum via mass spectroscopy revealed potential prognostic markers of both severity and outcome [10,12]. Interestingly, metabolomics may also predict antiviral drug efficacy in COVID-19 [13], and metabolomic analysis of patients' exhaled air can identify patients with COVID-19 in acute respiratory distress syndrome. NMR-based metabolomic profiling of blood samples has been already used to monitor COVID-19 patients' response to tocilizumab [14].

We focused herein on the dynamics of metabolomic changes in blood plasma at three successive time points during the first week of COVID-19 patient hospitalisation, with patients divided into two groups: (i) those with a positive outcome (survivors) and (ii) those with a negative outcome (non-survivors or obviously worsening condition). Hospitalised COVID-19 patients with clinically proven moderate-to-severe pneumonia with acute hypoxemic respiratory failure were included. We were interested to explore the metabolic changes in blood plasma that could be associated with immune cell response, as well as with energy metabolism, in comparison to control subjects representing a sample of the normal population, without any acute or chronic inflammatory or pulmonary diseases. Secondarily, it was of interest as to whether there are metabolomic features in blood plasma that could predict patient outcome, at which time point are they recognisable, and to what extent. Complementary to testing significant changes, we also employed a discriminatory algorithm in the search for metabolites that could serve alone or in combination as plasma biomarkers.

2. Materials and Methods

2.1. Subjects

Altogether, 53 patients with PCR-confirmed SARS-CoV-2 were included in the study. Patients were admitted to the Clinic of Pneumology and Phthisiology, Martin University Hospital, Slovakia, due to chest X-ray/CT signs of bilateral pneumonia and acute hypoxemic respiratory failure requiring oxygen supplementation (oxygen saturation at <94% in room air). In general, patients presented with typical symptoms of COVID-19: fever, cough, dyspnoea, weakness, fatigue, myalgia and arthralgia, loss of smell and taste, and loss of appetite. Some patients suffered from gastrointestinal symptoms (diarrhoea) as well. Laboratory results on admission showed increased inflammatory markers (CRP, IL-6, ferritin, fibrinogen) and hypoxemic respiratory failure, and changes in differential blood count included leucocytosis, lymphopenia, neutrophilia, and eosinopenia in most patients.

During the study, patients received either standard hospital enteral nutrition or a diabetic diet (patients with diabetes). Patients incapable of oral food intake received the equivalent for enteral nutrition via nasogastric tube. None of the included patients had percutaneous endoscopic gastrostomy/jejunostomy. Neither nutritional supplementation nor parenteral nutrition was administered. When necessary, but only sporadically, patients received crystalloid solutions to treat dehydration or mineral imbalance.

Oxygen was administered via nasal cannula, face mask, or face mask with a rebreathing bag with flow adjusted to achieve target oxygen saturation of 94%. Seven patients required high-flow nasal oxygen therapy (HFNO), and in case of hypoxemic–hypercapnic respiratory failure, three received non-invasive ventilation (NIV). In patients with severe and critical clinical condition requiring a very high flow of oxygen, saturation of 90% was considered sufficient. None of the included patients received mechanical ventilation during sample collection; however, two patients were later intubated and mechanically ventilated. Apart from oxygen supply, patients were treated with dexamethasone (all patients, dose of 6 mg/day for a duration of 10 days); antivirals (remdesivir or favipiravir if eligible according to local guidelines—duration of symptoms less than 7 days), n = 17; antibiotics (in case of bacterial superinfection or its suspicion), n = 53; LMWH, n = 49; vitamins: vitamin C, n = 17, vitamin D, n = 19; zinc, n = 14; and betaglucans, n = 44.

Patients were divided into two subgroups: Group A (n = 34) contained patients with a positive outcome (survivors), while Group B (n = 19) contained patients with a negative outcome, i.e., patients with a worsening condition during the sampling period, or those who died (10 were dead at the time of manuscript preparation). All known patient comorbidities at the time of study enrolment are listed in Table 1. To assess the patients' condition, the determining criterion was the need for increasing/decreasing oxygen flow or switch to HFNO, NIV, or mechanical ventilation to achieve target oxygen saturation, together with clinical evaluation and known clinical outcome. Due to various causes such as hospital discharge before Day 7, death, or even patient disagreement with other blood collections, the number of samples on Day 3 or Day 7 is slightly reduced. All details about subjects included in the study are summarised in Table 1.

Table 1. Characteristics of patients included in the study.

Parameter: Median (IQR)	Group A	Group B
size	34	19
age, years	65 (21)	71 (16)
gender	15 female	8 female
number of samples Day 1	34	19
number of samples Day 3	31	16
number of samples Day 7	26	10
oxygen	34	19
HFNO	-	7
NIV	1 *	2
smoker	2	2
non smoker	23	10
ex-smoker	5	5
smoking not known	4	2
chronic obstructive pulmonary disease	2	4
obesity	11	11
hypertension	22	13
asthma	2	-
kidney disease	4	4
ischemic heart disease	9	9
diabetes	14	7
cancer	1	2
cancer history	1	3
thyroid disease	3	1
liver cirrhosis	-	1
rheumatoid arthritis	1	3
stroke history	1	1
acute stroke	1	-
sarcodiosis	1	-

* Patient with chronic hypoxemic–hypercapnic respiratory failure due to COPD on home NIV (non-invasive ventilation) with LTOT (long term oxygen therapy).

As controls, plasma samples from age- and gender-matched subjects without any acute or chronic inflammatory diseases, any type of respiratory failure, or any pulmonary diseases, regardless of common highly age-related conditions (hypertensia, obesity, and others in the representative sample of the population) were used, representing a 'sample of the normal population', collected in a fasting state without any additional criteria. Included were 55 subjects: median age 64, IQR 18, female n = 25.

2.2. Sample Preparation

Blood was collected in EDTA-coated tubes, in the fasting state, after the first night in the hospital (Day 1) and then 2 and 6 days later (Day 3 and Day 7). Within 1 h after collection, blood was centrifuged to plasma at 4 °C, at 2000 rpm, for 20 min and stored at −80 °C until use. Plasma denaturation was carried out according to Gowda et al. [15]:

600 μL of methanol was added to 300 μL of blood plasma. The mixture was briefly vortexed and frozen at −24 °C for 20 min. After subsequent centrifugation at 14,000 rpm for 15 min, 700 μL of supernatant was taken, dried out, and stored at −24 °C. Before NMR measurement, the dried matter was mixed with 100 μL of stock solution (consisting of: phosphate buffer 200 mM pH 7.4 and 0.30 mM TSP-d_4 (trimethylsilylpropionic acid -d_4) as a chemical shift reference in deuterated water) and 500 μL of deuterated water. Finally, 550 μL of the final mixture was transferred into a 5 mm NMR tube.

2.3. NMR Measurement

NMR data were acquired on a 600 MHz Avance III NMR spectrometer from Bruker, Germany, equipped with a TCI CryoProbe at T = 310 K. Initial settings (basal shimming, receiver gain, and water suppression frequency) were performed on an independent sample and adopted for measurements. After preparation, samples were stored in a Sample Jet automatic machine, cooled at approximately 5 °C. Before measurement, each sample was preheated to 310 K for 5 min. An exponential noise filter was used to introduce 0.3 Hz line broadening before Fourier transform. All data were zero-filled. Samples were randomly ordered for acquisition.

We modified standard profiling protocols from Bruker as follows: denaturised plasma: noesy with presaturation (noesygppr1d): FID size 64k, dummy scans 4, number of scans 64, spectral width 20.4750 ppm; profiling cpmg (cpmgpr1d, L4 = 126, d20 = 3ms): number of scans 64, spectral width 20.4750 ppm. For 15 randomly chosen samples, 2D spectra were measured: cosy with presaturation (cosygpprqf): FID size 4k, dummy scans 8, number of scans 16, spectral width 16.0125 ppm; homonuclear J-resolved (jresgpprqf): FID size 8k, dummy scans 16, number of scans 32. Samples were randomly ordered for acquisition. For denaturised plasma samples, we kept the half-width of the TSP-d_4 signal under 1.0 Hz. All experiments were conducted with a relaxation delay of 4 s.

2.4. Data Processing

Spectra were solved using the human metabolomic database (www.hmda.ca, accessed on 23 March 2021) [16], chenomics software free trial version, internal metabolite database, and research in the metabolomic literature [15]. The proton NMR chemical shifts are reported relative to the TSP-d_4 signal assigned a chemical shift of 0.000 ppm. The peak multiplicities were confirmed in J-resolved spectra, and homonuclear cross peaks were confirmed in 2D cosy spectra. Peak assignments are listed in Table 2.

All spectra were binned to bins of size 0.001 ppm. No normalisation method was applied to the data. Then, the intensities of selected bins were summed only for spectra subregions with only one metabolite assigned or minimally affected by other co-metabolites. Metabolites showing weak intensive peaks or strong peak overlap were excluded from the evaluation. The obtained values were used as relative concentrations of particular metabolites.

Besides principal component analysis (PCA) and partial least squares discriminant analysis (PLS-DA), we applied the random forest (RF) discriminatory algorithm on the data. We ran nonparametric ANOVA (Kruskal–Wallis) and the nonparametric Mann–Whitney U-test to test significance. For data processing and analyses, we used the online tool metaboanalyst 5.0 [17], Origin Pro 2019, PASW Statistics software, and Matlab 2018b.

Table 2. Chemical shifts (in ppm), J couplings (in Hz), and multiplicities (s, singlet; d, doublet; t, triplet; q, quartet; m, multiplet; dd, doublet of doublets; dq, doublet of quartets) for the pool of metabolites identified in blood plasma. Signals marked with # were not suitable for quantitative analyses.

Metabolite	NMR Peak Assignment, Confirmed by Jres and Cosy
lactate	1.33 (d; J = 7.0), 4.12 (q; J = 7.0)
glutamine	2.12 (m), 2.15 (m), 2.44 (m), 2.48 (m), 3.77 (dd)
isoleucine	0.94 (t; J = 7.5), 1.01 (d; J = 7.0), 3.68 (d; J = 4.2)
leucine	0.96 (d; J = 6.2), 0.97 (d; J = 6.1), 1.68 (m), 1.72 (m), 1.75(m)
phenylalanine	3.13 (m), 3.28 (m), 7.34 (d; J = 7.5), 7.38 (t; J = 7.4), 7.44 (t)
tyrosine	3.05 (dd), 3.20 (dd), 3.93 (dd), 6.91 (d; J = 8.5), 7.20 (d; J = 8.5)
valine	0.99 (d; J = 7.1), 1.04 (d; J = 7.1), 2.27 (m), 3.61 (d; J = 4.4)
pyruvate	2.38 (s)
citrate	2.54 (d), 2.67 (d)
acetate	1.92 (s)
alanine	1.48 (d; J = 7.30), 3.78 (q)
glucose	3.23 (m), 3.40 (m), 3.46 (m), 3.52 (dd), 3.78 (m), 3.82 (m), 3.89 (dd), 4.64 (d), 5.23 (d)
3-hydroxybutyrate	1.20 (d; J = 6.23 Hz), 2.31 (m), 2.41 (m), 4.16 (m)
creatine	3.04 (s), 3.94 (s)
lysine	1.33 (d), 3.58 (d; J = 4.9), 4.25 (m)
2-oxoisocapronate (2-ketoleucine)	0.94 (d; J = 6.6), 2.11 (m), 2.61 (d; J = 7.0)
α-ketoisovalerate (2-ketovaline)	1.11 (d; J = 7.1), 3.01(dq)
3-methyl-2-oxo-valerate (2-ketoisoleucine)	0.90 (t; J = 7.5), 1.10 (d; J = 6.7)
lipoprotein fraction	0.82–0.93 (m), 1.20–1.37 (m)
# creatinine	3.05 (s), 4.07 (s)
# histidine	7.07 (s), 7.80 (s)
# proline	1.46 (m), 1.50 (m), 1.73 (m), 1.89 (m), 1.93 (m), 3.03 (t; J = 7.6)
# threonine	1.34 (d), 3.56 (d;J = 4.9),4.26 (m)
# tryptophan	7.21 (t), 7.30 (t), 7.33 (s), 7.56 (d), 7.74 (d; J = 8.0)

3. Results

Altogether, 24 metabolites were identified in denatured plasma in both patients and healthy subjects, where the signals from 19 compounds were sufficient for quantitative evaluation (Table 2). Further in the text, we use the trivial names of 2-ketoacids derived from leucine, isoleucine, and valine (IUPAC names are in Table 2). Besides molecular metabolites, we also evaluated the lipoprotein fraction, which, as described by Liu et al., contains very-low-density lipoproteins (VLDL), low-density lipoproteins (LDL), and high-density lipoproteins (HDL), including up to one-third of triacylglycerides [18]. For multivariate analyses, we used the relative concentrations of plasma metabolites (expressed as the integral of a particular spectral region) as an input in order to target biologically informative value. We avoided feeding the algorithms with binned NMR spectra as is common in metabolomic studies, since there may be regions of NMR spectra marked as important that are not straightforward and unambiguously related to biological relevance.

Firstly, the data of all patients were analysed (Group A and Group B together) on Day 1 against controls by PCA and PLS-DA (Figure 1). In contrast to patients, controls were relatively clustered together. The loading values were the highest for glucose, 3-hydroxybutyrate, and leucine in PC1 and alanine, lactate, and glutamine in PC2. The situation was very similar after the PLS-DA run. The 10-fold cross-validated PLS-DA algorithm performed with accuracy of 0.954, R2 of 0.7926, and Q2 of 0.6749 for eight components. The variables with the highest VIP scores were: glucose, 3-hydroxybutyrate, alanine, leucine, valine, and glutamine (performance measured in accuracy). The incorporation of additional variables did not improve the performance.

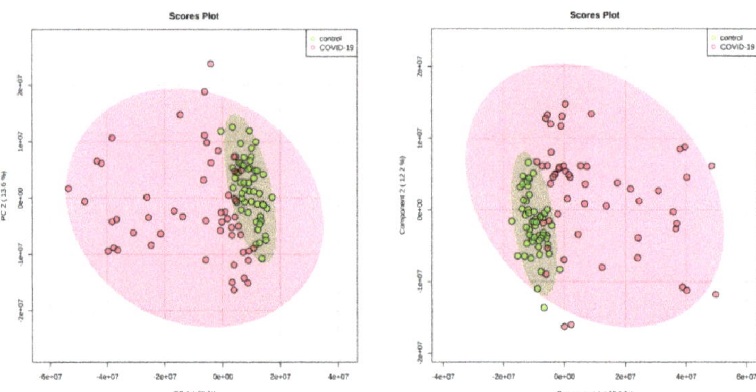

Figure 1. PCA (**left**) and PLD-DA analyses (**right**) of the system: patients in the hospital on Day 1 versus controls; algorithms were fed by the relative concentrations of plasma metabolites, and analyses were run in metaboanalyst [16].

The PCA and PLS-DA analyses of the ternary system comprising Group A and Group B on Day 1 and the controls showed a very similar result to those for the previous binary system, where the patients were clustered together relatively well and patients were scattered among themselves without obvious differentiation between patient groups (results shown in Figure S1 in the supplement). PLS-DA analyses were further used to differentiate patient data on a given day. The results from these analyses can be summarised as follows (the best result, performance measured in accuracy): Day 1, accuracy of 0.73, R2 of 0.138 (one component); Day 3, accuracy of 0.76, R2 of 0.3905 (five components); and Day 7, accuracy of 0.72, R2 of 0.387 (four components). In all cases, Q2 values were negative, which suggests an overfitted model.

In the next step, we employed the random forest (RF) discriminatory algorithm to obtain a more realistic estimation of the discriminatory power of the system since RF is relatively robust to overfitting and outliers [19]. The RF algorithm used included cross-validation via balanced subsampling. It worked with two-thirds of the data for training and the rest for testing for regression, and about 70% of the data for training and the rest for testing during classification to overcome the negative aspects of training and testing on the same data. This approach partially substitutes the validation on an independent data set. As input variables also for this algorithm, we used relative concentrations of metabolites in plasma expressed by the spectral integrals of particular NMR regions. In the case of highly correlating predictors, RF may label some of them as unimportant, so RF was launched 10 times. Within the RF re-runs, metabolites slightly permuted in the importance order. As an output from these analyses, receiver operating characteristic curve (ROC) curves were created. The ROC is defined only for binary systems, and it is created by plotting the true-positive rate against the false-positive rate at various threshold settings. An important output is the area under the curve (AUC), which represents ranking quality. The AUC of a ranking is 1 (the maximum AUC value) when all samples are truly assigned into the groups. An AUC of 0.5 is equivalent to randomly classifying subjects as either positive or negative (i.e., the classifier is of no practical utility) [20]. We ran RF discriminatory analyses for the systems of patients versus controls, Group A versus controls, Group B versus controls, and Group A versus Group B on Days 1, 3, and 7. The results of RF classifications are summarised in Table 3.

Table 3. Outputs from random forest discriminatory analyses for selected systems.

System	OOB Error (5 Variables)	AUC	Number of Variables	Metabolites in Importance Order
All Patients Day 1/Controls	3/108	0.984 0.995	2 5	3-hydroxybutyrate, phenylalanine, Phe/Tyr ratio, acetate, glucose
Group A Day 1/Controls	1/89	0.977 0.996	2 5	3-hydroxybutyrate, phenylalanine, glucose, Phe/Tyr ratio, acetate,
Group B Day 1/Controls	1/74	0.972 0.991	2 5	phenylalanine, 3-hydroxybutyrate, Phe/Tyr ratio, acetate, glutamine or glucose
Group A/Group B Day1	-	0.568 0.674	2 5	AUC value too low
Group A/Group B Day 3	-	0.754 0.783	2 5	alanine, lysine, glutamine, Phe/Tyr ratio, phenylalanine
Group A/Group B Day 7	-	0.487 0.503	2 5	AUC value too low

For significance testing among relative concentrations of plasma metabolites in patients against controls and patients' dynamic data, we used nonparametric ANOVA, known as the Kruskal–Wallis test. Due to the relatively low sample sizes, we continued with nonparametric testing via the Mann–Whitney U-test for the combination of binary data sets. The details are listed in Table 4. The Phe/Tyr ratio was also used as one variable. As the threshold to claim significance, the p-value was set to 0.05, as established. In the discussion, we did not strictly adhere to p-values, but we focused rather on the data behaviour visualised in the box plots.

Table 4. Results from statistical tests; *p*-value derived from nonparametric ANOVA and Mann–Whitney U-test.

Metabolite	Nonparametric ANOVA (Kruskal–Wallis)			Mann–Whitney U-test, Only Significant Changes ($p < 0.05$) are Listed, Arrows Indicate the Direction of Change		
	chi. Squared	*p*-Value	FDR *p*-Value Adjusted	Group A Against Controls	Group B Against Controls	Group A Against GroupB
glucose	80	3.9×10^{-15}	1.4×10^{-14}	Day1↑, Day3↑	Day1↑, Day3↑, Day7↑	
3-OH-butyrate	130	1.1×10^{-24}	1.2×10^{-23}	Day1↑, Day3↑, Day7↑	Day1↑, Day3↑, Day7↑	Day7, A < B
citrate	77	1.2×10^{-14}	3.2×10^{-14}	Day1↓, Day3↓, Day7↓	Day1↓, Day3↓, Day7↓	
leucine	39	7.6×10^{-7}	1.5×10^{-6}	Day1↑, Day3↑, Day7↑	Day1↑, Day3↑, Day7↑	
isoleucine	31	2.8×10^{-5}	4.9×10^{-5}	Day1↑, Day3↑, Day7↑	Day1↑, Day3↑, Day7↑	
valine	13	0.040	0.047	Day1↑, Day3↑, Day7↑	Day1↑, Day3↑, Day7↑	
ketoleucine	25	3.7×10^{-4}	5.5×10^{-4}	Day1↑, Day3↓, Day7↓		Day1, A > B

Table 4. *Cont.*

Metabolite	Nonparametric ANOVA (Kruskal–Wallis)			Mann–Whitney U-test, Only Significant Changes ($p < 0.05$) are Listed, Arrows Indicate the Direction of Change		
	chi. Squared	p-Value	FDR p-Value Adjusted	Group A Against Controls	Group B Against Controls	Group A Against GroupB
ketoisoleucine	19	0.0042	0.0059	Day1↑, Day3↓, Day7↓		
ketovaline	17	0.0094	0.012	Day1↑, Day3↓, Day7↓		Day1, A > B
creatine	64	8.5×10^{-12}	2.0×10^{-11}	Day3↑, Day7↑	Day3↑, Day7↑	
alanine	49	8.5×10^{-9}	1.8×10^{-8}	Day1↓	Day1↓, Day3↓, Day7↓	Day3, Day7, A > B
glutamine	29	6.1×10^{-5}	9.8×10^{-5}	Day1↓	Day1↓,Day3↓	
phenylalanine	120	1.2×10^{-23}	8.5×10^{-23}	Day1↑, Day3↑, Day7↑	Day1↑, Day3↑, Day7↑	
Phe/Tyr ratio	85	3.3×10^{-16}	1.4×10^{-15}	Day1↑, Day3↑, Day7↑	Day1↑, Day3↑, Day7↑	Day1, A < B
lipoproteins	150	7.8×10^{-31}	1.6×10^{-29}	Day1↓, Day3↓, Day7↓	Day1↓, Day3↓, Day7↓	
acetate	100.	2.1×10^{-19}	1.1×10^{-18}	Day1↓, Day3↓, Day7↓	Day1↓, Day3↓, Day7↓	
lysine	79	6.6×10^{-15}	2.0×10^{-14}	Day1↑, Day3↑, Day7↑	Day7↑	

4. Discussion

4.1. Discriminatory Analyses

PCA and PLS-DA analyses are well-established tools when evaluating multidimensional data. PCA analysis serves rather as a 2D visualisation of data sets indicating group proximity. PLS-DA includes a discriminatory algorithm and may be used also to differentiate among groups. PCA analysis of the patient data collected on Day 1 against controls showed controls clustered together, whilst patients were scattered in 2D space. This suggests the great data variability in patient samples, which was more or less confirmed by PLS-DA. As PLS-DA is known to overfit the data [19], for biomarker discovery, we employed a cross-validated RF algorithm. As an output, the ROC curve was created. For the system of patients on Day 1 and controls, RF performed very well with an AUC of 0.995 for five variables with an out-of-bag error of 3/108. The variables Phe/Tyr ratio, phenylalanine, 3-hydroxybutyrate, acetate, and glucose were of the highest importance. The corresponding ROC curve is shown in Figure 2.

Very similar performance—almost ideal discrimination—was achieved for the systems of Group A on Day 1 against controls and Group B on Day 1 against controls (details in Table 3). The five metabolites of the highest importance were identical to those before: phenylalanine, Phe/Tyr ratio, acetate, 3-hydroxybutyrate, glucose, permuted with glutamine, and proline.

The possibility to discriminate between acute COVID-19 patients and healthy controls has been proven in previous studies [7,10,11]. These studies covered another spectrum of metabolites evaluated by different analytical tools as NMR spectroscopy. Here, we also note that metabolites that were marked as the most important in the discrimination algorithm may not be specific to COVID-19 disease, since as discussed in the next text, they are generally related to inflammation, immune response, and energy metabolism.

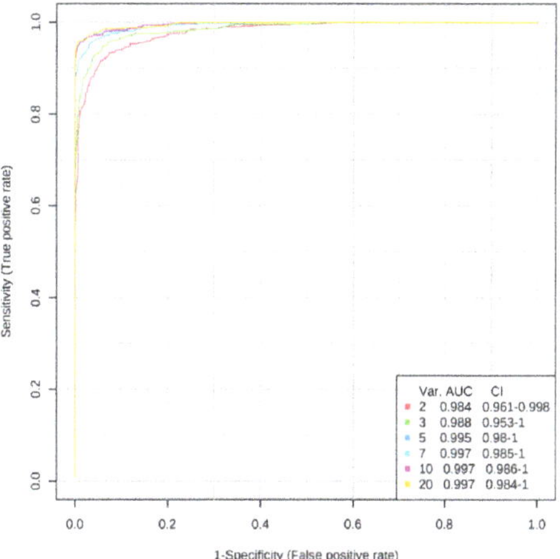

Figure 2. ROC curve with AUC values for systems of COVID-19 patients on Day 1 vs. controls, determined by random forest algorithm with relative concentrations of metabolites in blood plasma as input variables; analysis run in metaboanalyst [16].

It was of interest to see whether there are any metabolites in blood plasma that could serve as potential predictors of disease progress/outcome. We ran RF discrimination for binary systems of patients' groups on collection days. Here the performance was weaker, with AUC values of 0.67 on Day 1 and 0.78 on Day 3 for common, permuting variables: Phe/Tyr ratio, alanine, lysine, glutamine, leucine, and phenylalanine. A further increase in the number of variables did not improve the performance of the discrimination analysis. For the data set of Group A versus Group B on Day 7, the system did not show any discriminatory potential, with an AUC value of 0.503, in other words, the classification was not relevant. Based on this, the biochemical changes observed were rather indicative, not defining unambiguous biomarkers for patient outcome.

4.2. Metabolomic Changes

Patients hospitalised due to a severe course of COVID-19 showed a significantly increased glucose level on Day 1. All patients were equally treated over the whole time period with dexamethasone, which is known to impair glucose metabolism [21] via the stimulation of gluconeogenesis from amino acids released from muscles, and even one dose of 10 mg dexamethasone may lead to a temporarily increased blood glucose level [22]. The hyperglycaemia in COVID-19 patients treated with dexamethasone is presumably caused by 'triple insult': dexamethasone-induced impaired glucose metabolism, COVID-19-induced insulin resistance, and COVID-19 impaired insulin production [23]. Prolonged uncontrolled hyperglycaemia, regardless of diabetes mellitus, seems to be important in the pathogenesis of COVID-19 [24]. In our study, the hyperglycaemia normalised in Group A, but not in patients with unfavourable outcome included in Group B (Figure 3). This observed result is in line with general knowledge that hyperglycaemia is an unfavourable state in many clinical conditions, i.a., in severe inflammation [25], and is one of the important risk factors of COVID-19 disease progression [26]. The plasma levels of glycolytic products pyruvate and, eventually, lactate were not significantly changed in any group of patients. The relative plasma level of alanine, a metabolite that contributes significantly to liver gluconeogenesis, was decreased on Day 1 in both groups but normalised in patients with a positive outcome on Days 3 and 7; however, it stayed decreased in Group B (figure not shown).

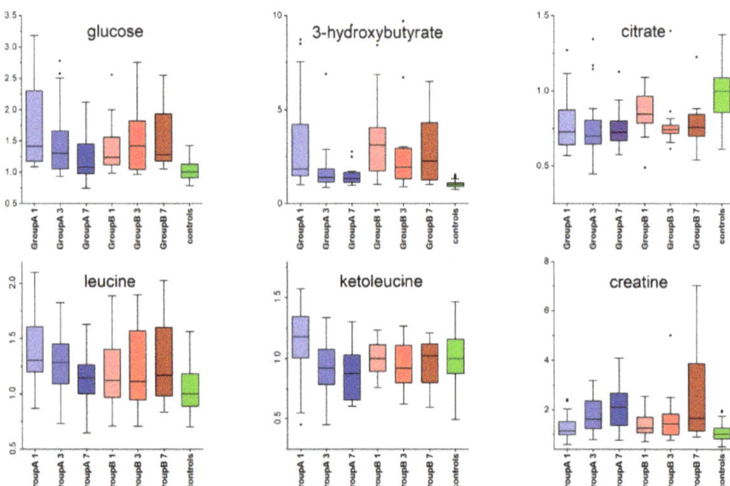

Figure 3. Relative concentrations of selected metabolites in blood plasma for patient Groups A and B on Day 1, Day 3, and Day 7. Values are relativised to the median of controls set to 1.

In the blood plasma of COVID-19 patients, we observed a significantly increased level of 3-hydroxybutyrate, a ketone bodies representative. Besides serving as an energy source for the brain, heart, and skeletal muscle, ketone bodies play pivotal roles as signalling mediators, drivers of protein post-translational modification, and modulators of inflammation and oxidative stress [27]. 3-hydroxybutyrate exerts a predominantly anti-inflammatory response [28–30], but can also be pro-inflammatory [31]. A recent study on COVID-19 patients already showed dysbalance in ketone bodies [32]. In our study, the initially increased plasma level of 3-hydroxybutyrate decreased over Day 3 and Day 7 in Group A, but it stayed at an elevated level in Group B on the third and seventh days (Figure 3). Interestingly, the glucose level in this patient group also remained high. As we did not analyse the level of C peptide as a representative of the insulin level, we can hypothesise that the proposed glucose resistance or insufficient glucose utilisation is compensated by ketone bodies. The increase in the 3-hydroxybutyrate level in COVID-19 patients is accompanied by a decreased amount of lipoprotein fraction in blood plasma in patients suffering from COVID-19, containing up to one-third of triacylglycerides [18] as one of the additional substrates for ketone body synthesis (boxplot not shown).

We observed a decreased citrate level in the blood plasma in COVID-19 patients, suggesting alteration of the TCA cycle (Figure 3), similar to the results of a recent study by Pang et al. [11]. Besides including α-ketoglutarate, an essential substrate for endogenous glutamate/glutamine synthesis, there is evidence that TCA cycle intermediates also have an epigenetic impact by influencing DNA and histone methylation, including immune cells [33]. Further, the metabolite creatine, a part of muscle energy metabolism, was significantly increased in the blood plasma of COVID-19 patients compared with controls in both groups, rising with the time of hospitalisation (Figure 3). Patients forced to lie in bed for a sustained period lack spontaneous movement utilising muscle energy, which is probably the reason for the increase of plasma creatine.

BCAAs (branched chain amino acids), including leucine, isoleucine, and valine, share a common pattern of extrahepatic metabolism, and their relative plasma concentrations were represented similarly in both groups of patients. In Figure 3, we show only the dynamics of leucine since isoleucine and valine behaved almost identically. As a representative of ketoacids derived from BCAAs, we show only the course of ketoleucine, as the dynamics was repeated for the other two ketoacids: ketovaline and ketoisoleucine. Increased leucine in COVID-19 patients was reported by Dierckx et al. [34]. There is an established association between elevated circulating BCAAs and their deleterious effects, as their increased concentration may promote oxidative stress and inflammation [35], having also a neurological impact [36,37]. By monitoring dynamic changes for two different patient subgroups, we observed that initially increased plasma levels of BCAAs in both groups slowly decreased in Group A but not in Group B (Figure 3). Interestingly, the mean values of BCAAs in Group B obviously follow the course of the plasma glucose levels. The increase of BCAAs at time of impaired glycolysis and increased use of fatty acids were very recently discussed in a comprehensive review by Holecek [38], showing the important role of BCAAs in energy metabolism.

Taking the above discussed results together, severe inflammation induced by COVID-19 caused changes in energy metabolism, where we observed increased blood glucose that implies lowered glucose utilisation (the influence of dexamethasone treatment cannot be omitted). In balance, the body, including immune cells, uses ketone bodies (observed increased 3-hydroxybutyrate together with decreased triacylglycerides) as an energy source, and, alternatively also amino acids released by accelerated protein catabolism (increased levels of essential amino acids BCAAs). Interestingly, although all patients in both groups received the dexamethasone treatment during the follow-up period, the above mentioned changes normalised only in patients with a positive outcome; however, they persisted in patients with a negative outcome (more than half of them had died at the time of writing). This course was independent of the patients' diet (Figure S2 in Supplement).

In acute inflammatory conditions, the demand on glutamine increases [39] which may lead to its plasma decrease if the endogenous synthesis of glutamine does not fulfil the requirements of the body [39]. Glutamine serves besides others as a fuel for immune cells—lymphocytes, neutrophils, and macrophages [39–42]—and plays a crucial role in cytokine production [42]. In our study, we noticed a decrease in the glutamine plasma level in COVID-19 patients on Day 1, observed to a lower extent in Group A, which is in accordance with another study where glutamine deficiency may have contributed to disease severity [43]. The glutamine plasma level normalised in both groups, but this was faster in Group A (Figure 4). On Day 7, both groups of patients showed plasma glutamine levels very similar to the level in control subjects, where probably the balance between glutamine production and utilisation had stabilised (Figure 4). Accelerated spontaneous stabilisation of glutamine levels in patients with better outcome supports the results from another study, where the administration of glutamine in the early period of infection suggested a shortened hospital stay and decreased the need for ICU stay [40].

Another significant metabolic parameter associated with immune activation and inflammation is the Phe/Tyr ratio [44,45]. Perturbations in phenylalanine and tyrosine biosynthesis were recognised in SARS-CoV-2 patients by Barberis et al. [46]. In our study, both groups showed initially increased plasma phenylalanine levels, as observed in another study [34], and the level tended to decrease in Group A but not in Group B (Figure 4). The plasma tyrosine level did not show any substantial change. The Phe/Tyr ratio was calculated by dividing the relative concentrations of both metabolites. The obtained value is only the relative ratio, but for comparison, it has the same informative value. The Phe/Tyr ratio was increased in both groups, obviously higher in patients with unfavourable outcome, where a course towards control levels was slowed down in Group B against Group A (Figure 4). Positive relationships between the Phe/Tyr ratio and immune activation markers have been described earlier in several papers [44,45]. It was suggested that suppression of body inflammation can, to a certain extent, improve abnormalities in Phe metabolism within associated neuropsychiatric symptoms [44], among which, e.g., depression and fatigue are some of the most recognised post-COVID-19 difficulties [47].

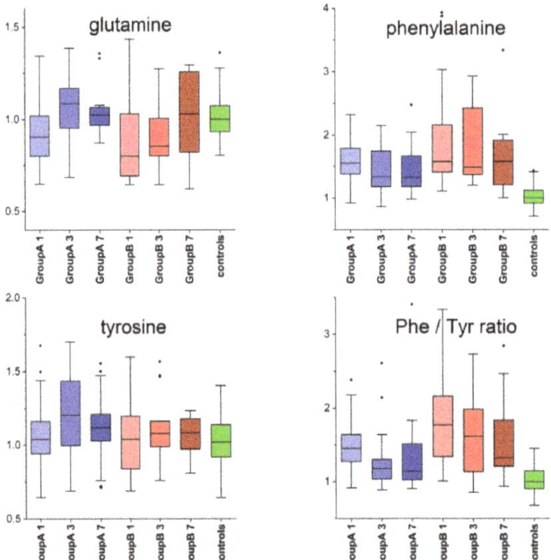

Figure 4. Relative concentrations of selected metabolites related to immunity in blood plasma for patient Groups A and B on Day 1, Day 3, and Day 7. Values are relativised to the median of controls given a value of 1.

5. Conclusions

Metabolomic changes in blood plasma analysed by NMR in patients suffering COVID-19 were strong enough to obtain almost ideal discrimination from controls, where the ROC derived from random forest showed an AUC of 0.995 for the variables 3-hydroxybutyrate, phenylalanine, acetate, glucose, and Phe/Tyr ratio. The inflammation by COVID-19 caused changes in the body's energy metabolism, where we observed increased blood glucose that implies lowered glucose utilisation, balanced with increased production of 3-hydroxybutyrate as an alternative energy source. Besides that, increased essential BCAAs are a sign of accelerated protein catabolism, offering a further energy source. Interestingly, although all COVID-19-positive patients received dexamethasone treatment during the follow-up period, the above mentioned changes (increased glucose, 3-hydroxybutyrate, and BCAAs levels in blood plasma) normalised only in patients with positive outcome by the seventh day; however, they persisted for over one week in patients with negative outcome (more than half of them had died at the time of writing). Further, patients suffering COVID-19 showed decreased plasma glutamine that normalised faster in patients with a positive outcome. With the length of hospital stay, plasma levels of creatine increased in patients in both groups. Increased Phe/Tyr ratio, which is closely connected with neuropsychiatric morbidities, often reported as post-COVID-19 symptoms, was more pronounced in patients with a negative outcome. Based on our results, the ability of patients to normalise energy metabolism seems to be one of the key factors determining the disease progression. This trend was observed independently of patient diet, which differed with respect to diabetic condition. This study documents evident differences in the course of the metabolomic response to COVID-19 in relation to patient outcome. However, the described changes may not be unique for COVID-19 since they reflect generalised immune response and alterations in body energy metabolism as well. The presented results may serve towards complementary personalised pharmacological and nutritional support in order to improve patient outcomes in severe inflammatory conditions.

Supplementary Materials: The following are available online at https://www.mdpi.com/article/10.3390/app11094231/s1, Figure S1: PCA (left) and PLD-DA analyses (right) of the system of patients divided into subgroups Group A and Group B on Day 1 versus controls; algorithms were fed relative concentrations of plasma metabolites, and analyses were run in metaboanalyst [16]. Figure S2. The relative changes in two metabolites closely related to energy metabolism—glucose and 3-hydroxybutyrate—where both Groups A and B were divided into subgroups according to patient diet (according to presence of diabetes) on Days 1, 3, and 7 after hospital arrival; not dia = non-diabetic patients on a normal diet, dia = diabetic patients on a diabetic diet. Values were relativized to median of controls set to 1.

Author Contributions: Conceptualization, E.B., A.B., R.V., V.N. and J.L.; methodology, E.B.; formal analysis, E.B.; investigation, E.B., Z.D. and J.L.; data curation, E.B., A.B., R.V., Z.D. and J.L.; writing—original draft preparation, E.B..; writing—review and editing, A.B., E.H., V.N. and J.L.; supervision, J.L.; funding acquisition, J.L. All authors have read and agreed to the published version of the manuscript.

Funding: This publication has been produced with the support of the Integrated Infrastructure Operational Program for the project: New possibilities for laboratory diagnostics and massive screening of SARS-Cov-2 and identification of mechanisms of virus behaviour in human body, ITMS: 313011AUA4, co-financed by the European Regional Development Fund and grant VEGA No. 230/20 and APVV 15/0107

Institutional Review Board Statement: This study was approved by the Ethics Committee of the Jessenius Faculty of Medicine in Martin (registered under IRB00005636 at Office for Human Research Protection, U.S. Department of Health and Human Services) under the code EK82/2020.

Informed Consent Statement: Informed consent was obtained from all subjects involved in the study.

Data Availability Statement: The NMR spectra or evaluated data used in this study are available on request from the author: eva.baranovicova@uniba.sk.

Acknowledgments: We would like to thank the medical staff from the Clinic of Pneumology and Phthisiology, Martin University Hospital, Slovakia, for their helpfulness during sample collection. This publication was produced with the support of the Integrated Infrastructure Operational Program for the project: New possibilities for laboratory diagnostics and massive screening of SARS-CoV-2 and identification of mechanisms of virus behaviour in human body, ITMS: 313011AUA4, co-financed by the European Regional Development Fund and grant VEGA No. 230/20.

Conflicts of Interest: The authors declare no conflict of interest.

References

1. Wellen, K.E.; Hotamisligil, G.S. Inflammation, stress, and diabetes. *J. Clin. Investig.* **2005**, *115*, 1111–1119. [CrossRef]
2. Jung, J.; Zeng, H.; Horng, T. Metabolism as a guiding force for immunity. *Nat. Cell Biol.* **2019**, *21*, 85–93. [CrossRef]
3. McGaha, T.L.; Huang, L.; Lemos, H.; Metz, R.; Mautino, M.; Prendergast, G.C.; Mellor, A.L. Amino acid catabolism: A pivotal regulator of innate and adaptive immunity. *Immunol. Rev.* **2012**, *249*, 135–157. [CrossRef] [PubMed]
4. Jafar, N.; Edriss, H.; Nugent, K. The Effect of Short-Term Hyperglycemia on the Innate Immune System. *Am. J. Med Sci.* **2016**, *351*, 201–211. [CrossRef]
5. Berbudi, A.; Rahmadika, N.; Tjahjadi, A.I.; Ruslami, R. Type 2 Diabetes and its Impact on the Immune System. *Curr. Diabetes Rev.* **2020**, *16*, 442–449. [CrossRef] [PubMed]
6. Geerlings, S.E.; Hoepelman, A.I.M. Immune dysfunction in patients with diabetes mellitus (DM). *FEMS Immunol. Med. Microbiol.* **1999**, *26*, 259–265. [CrossRef] [PubMed]
7. Wu, D.; Shu, T.; Yang, X.; Song, J.-X.; Zhang, M.; Yao, C.; Liu, W.; Huang, M.; Yu, Y.; Yang, Q.; et al. Plasma metabolomic and lipidomic alterations associated with COVID-19. *Natl. Sci. Rev.* **2020**, *7*, 1157–1168. [CrossRef]
8. Song, J.-W.; Lam, S.M.; Fan, X.; Cao, W.-J.; Wang, S.-Y.; Tian, H.; Chua, G.H.; Zhang, C.; Meng, F.-P.; Xu, Z.; et al. Omics-Driven Systems Interrogation of Metabolic Dysregulation in COVID-19 Pathogenesis. *Cell Metab.* **2020**, *32*, 188–202. [CrossRef]
9. Blasco, H.; Bessy, C.; Plantier, L.; Lefevre, A.; Piver, E.; Bernard, L.; Marlet, J.; Stefic, K.; Bretagne, I.B.-D.; Cannet, P.; et al. The specific metabolome profiling of patients infected by SARS-COV-2 supports the key role of tryptophan-nicotinamide pathway and cytosine metabolism. *Sci. Rep.* **2020**, *10*, 1–12. [CrossRef]
10. Shen, B.; Yi, X.; Sun, Y.; Bi, X.; Du, J. Proteomic and metabolomic characterisation of COVID-19 patient sera. *Cell* **2020**, *182*, 59–72. [CrossRef] [PubMed]
11. Pang, Z.; Zhou, G.; Chong, J.; Xia, J. Comprehensive Meta-Analysis of COVID-19 Global Metabolomics Datasets. *Metabolites* **2021**, *11*, 44. [CrossRef]

12. Roberts, I.; Muelas, M.W.; Taylor, J.M.; Davison, A.S.; Xu, Y.; Grixti, J.M.; Gotts, N.; Sorokin, A.; Goodacre, R.; Kell, D.B. Untargeted metabolomics of COVID-19 patient serum reveals potential prognostic markers of both severity and outcome. *medRxiv* **2020**. [CrossRef]
13. Migaud, M.; Gandotra, S.; Chand, H.S.; Gillespie, M.N.; Thannickal, V.J.; Langley, R.J. Metabolomics to Predict Antiviral Drug Efficacy in COVID-19. *Am. J. Respir. Cell Mol. Biol.* **2020**, *63*, 396–398. [CrossRef]
14. Meoni, G.; Ghini, V.; Maggi, L.; Vignoli, A.; Mazzoni, A.; Salvati, L.; Capone, M.; Vanni, A.; Tenori, L.; Fontanari, P.; et al. Metabolomic/lipidomic profiling of COVID-19 and individual response to tocilizumab. *PLoS Pathog.* **2021**, *17*, e1009243. [CrossRef]
15. Gowda, G.A.N.; Gowda, Y.N.; Raftery, D. Expanding the Limits of Human Blood Metabolite Quantitation Using NMR Spectroscopy. *Anal. Chem.* **2015**, *87*, 706–715. [CrossRef]
16. Wishart, D.S.; Feunang, Y.D.; Marcu, A.; Guo, A.C.; Liang, K.; Vázquez-Fresno, R.; Sajed, T.; Johnson, D.; Allison, P.; Karu, N.; et al. HMDB 4.0: The human metabolome database for 2018. *Nucleic Acids Res.* **2018**, *46*, D608–D617. [CrossRef] [PubMed]
17. Xia, J.; Wishart, D.S. Using MetaboAnalyst 3.0 for Comprehensive Metabolomics Data Analysis. *Curr. Protoc. Bioinform.* **2016**, *55*, 14-10. [CrossRef]
18. Liu, M.; Tang, H.; Niholson, J.K.; Lindon, J.C. Use of 1H NMR-determined diffusion coefficients to characterise lipoprotein fractions in human blood plasma. *Magn. Reson. Chem.* **2002**, *40*, S83–S88. [CrossRef]
19. Gromski, P.S.; Muhamadali, H.; Ellis, D.I.; Xu, Y.; Correa, E.; Turner, M.L.; Goodacre, R. A tutorial review: Metabolomics and partial least squares-discriminant analysis—A marriage of convenience or a shotgun wedding. *Anal. Chim. Acta* **2015**, *879*, 10–23. [CrossRef]
20. Xia, J.; Broadhurst, D.I.; Wilson, M.; Wishart, D.S. Translational biomarker discovery in clinical metabolomics: An introductory tutorial. *Metabolomics* **2012**, *9*, 280–299. [CrossRef]
21. Tamez-Pérez, H.E.; Quintanilla-Flores, D.L.; Rodríguez-Gutiérrez, R.; González-González, J.G.; Tamez-Peña, A.L. Steroid hyperglycemia: Prevalence, early detection and therapeutic recommendations: A narrative review. *World J. Diabetes* **2015**, *6*, 1073–1078. [CrossRef]
22. Pasternak, J.J.; McGregor, D.G.; Lanier, W.L. Effect of Single-Dose Dexamethasone on Blood Glucose Concentration in Patients Undergoing Craniotomy. *J. Neurosurg. Anesthesiol.* **2004**, *16*, 122–125. [CrossRef]
23. Rayman, G.; Lumb, A.N.; Kennon, B.; Cottrell, C.; Nagi, D.; Page, E.; Voigt, D.; Courtney, H.C.; Atkins, H.; Higgins, K.; et al. Dexamethasone therapy in COVID-19 patients: Implications and guidance for the management of blood glucose in people with and without diabetes. *Diabet. Med.* **2021**, *38*, e14378. [CrossRef] [PubMed]
24. Brufsky, A. Hyperglycemia, hydroxychloroquine, and the COVID-19 pandemic. *J. Med Virol.* **2020**, *92*, 770–775. [CrossRef] [PubMed]
25. Butkowski, E.G.; Jelinek, H.F. Hyperglycaemia, oxidative stress and inflammatory markers. *Redox Rep.* **2016**, *22*, 257–264. [CrossRef]
26. Erener, S. Diabetes, infection risk and COVID-19. *Mol. Metab.* **2020**, *39*, 101044. [CrossRef]
27. Puchalska, P.; Crawford, P.A. Multi-dimensional Roles of Ketone Bodies in Fuel Metabolism, Signaling, and Therapeutics. *Cell Metab.* **2017**, *25*, 262–284. [CrossRef]
28. Fu, S.P.; Li, S.N.; Wang, J.F.; Li, Y.; Xie, S.S.; Xue, W.J.; Liu, H.M.; Huang, B.X.; Lv, Q.K.; Lei, L.C.; et al. BHBA suppresses LPS-induced inflammation in BV-2 cells by inhibiting NF-κB activation. *Mediat. Inflamm.* **2014**, *2014*, 983401. [CrossRef]
29. Rahman, M.; Muhammad, S.; Khan, M.A.; Chen, H.; Ridder, D.A.; Muller-Fielitz, H.; Pokorna, B.; Vollbrandt, T.; Stolting, I.; Nadrowitz, R.; et al. The Β-hydroxybutyrate receptor HCA2 activates a neuroprotective subset of macrophages. *Nat. Commun.* **2014**, *5*, 3944. [CrossRef] [PubMed]
30. Youm, Y.; Nguyen, K.Y.; Grant, R.W.; Goldberg, E.L.; Bodogai, M.; Kim, D.; D'agostino, D.; Planavsky, N.; Lupfer, C.; Kanneganti, T.D.; et al. The ketone metabolite beta-hydroxybutyrate blocks NLRP3 inflammasome-mediated inflammatory disease. *Nat. Med.* **2015**, *21*, 263–269. [CrossRef]
31. Shi, X.; Li, X.; Li, D.; Li, Y.; Song, Y.; Deng, Q.; Wang, J.; Zhang, Y.; Ding, H.; Yin, L.; et al. β-Hydroxybutyrate Activates the NF-κB Signaling Pathway to Promote the Expression of Pro-Inflammatory Factors in Calf Hepatocytes. *Cell Physiol. Biochem.* **2014**, *33*, 920–932. [CrossRef] [PubMed]
32. Li, J.; Wang, X.; Chen, J.; Zuo, X.; Zhang, H.; Deng, A. COVID -19 infection may cause ketosis and ketoacidosis. *Diabetes Obes. Metab.* **2020**, *22*, 1935–1941. [CrossRef] [PubMed]
33. Salminen, A.; Kauppinen, A.; Hiltunen, M.; Kaarniranta, K. Krebs cycle intermediates regulate DNA and histone methylation: Epigenetic impact on the aging process. *Ageing Res. Rev.* **2014**, *16*, 45–65. [CrossRef] [PubMed]
34. Dierckx, T.; van Elslande, J.; Salmela, H.; Decru, B.; Wauters, E.; Gunst, J.; Van, H.Y.; Wauters, J.; Stessel, B.; Vermeersch, P.; et al. The metabolic fingerprint of COVID-19 severity. *medRxiv* **2020**. [CrossRef]
35. Zhenyukh, O.; Civantos, E.; Ruiz-Ortega, M.; Sánchez, M.S.; Vázquez, C.; Peiró, C.; Egido, J.; Mas, S. High concentration of branched-chain amino acids promotes oxidative stress, inflammation and migration of human peripheral blood mononuclear cells via mTORC1 activation. *Free. Radic. Biol. Med.* **2017**, *104*, 165–177. [CrossRef]
36. HoleČek, M. Branched-chain amino acids in health and disease: Metabolism, alterations in blood plasma, and as supplements. *Nutr. Metab.* **2018**, *15*, 1–12. [CrossRef] [PubMed]

37. Larsson, S.C.; Markus, H.S. Branched-chain amino acids and Alzheimer's disease: A Mendelian randomisation analysis. *Sci. Rep.* **2017**, *7*, 13604. [CrossRef]
38. Holeček, M. Why Are Branched-Chain Amino Acids Increased in Starvation and Diabetes? *Nutrients* **2020**, *12*, 3087. [CrossRef]
39. Cruzat, V.; Rogero, M.M.; Keane, K.N.; Curi, R.; Newsholme, P. Glutamine: Metabolism and Immune Function, Supplementation and Clinical Translation. *Nutrients* **2018**, *10*, 1564. [CrossRef]
40. Cengiz, M.; Uysal, B.B.; Ikitimur, H.; Ozcan, E.; Islamoğlu, M.S.; Aktepe, E.; Yavuzer, H.; Yavuzer, S. Effect of oral l-Glutamine supplementation on Covid-19 treatment. *Clin. Nutr. Exp.* **2020**, *33*, 24–31. [CrossRef]
41. De Oliveira, D.C.; Lima, F.D.S.; Sartori, T.; Santos, A.C.A.; Rogero, M.M.; Fock, R.A. Glutamine metabolism and its effects on immune response: Molecular mechanism and gene expression. *Nutrients* **2016**, *41*, 14. [CrossRef]
42. Shah, A.M.; Wang, Z.; Ma, J. Glutamine Metabolism and Its Role in Immunity, a Comprehensive Review. *Animals* **2020**, *10*, 326. [CrossRef] [PubMed]
43. Abdelaal, M.A.; Abdelrahman, D.; Cengiz, M.; Yavuzer, H.; Yavuzer, S.; Bien, I.; Bhuva, P.; Pham, J.V.; Siu, R.; Tang, M.; et al. Actions of L-Glutamine vs. COVID-19 Suggest Additional Benefit in Sickle Cell Disease. *Blood* **2020**, *136*, 11–12. [CrossRef]
44. Murr, C.; Grammer, T.B.; Meinitzer, A.; Kleber, M.E.; März, W.; Fuchs, D. Immune Activation and Inflammation in Patients with Cardiovascular Disease Are Associated with Higher Phenylalanine to Tyrosine Ratios: The Ludwigshafen Risk and Cardiovascular Health Study. *J. Amino Acids* **2014**, *2014*, 1–6. [CrossRef]
45. Geisler, S.; Gostner, J.M.; Becker, K.; Ueberall, F.; Fuchs, D. Immune activation and inflammation increase the plasma phenylalanine-to-tyrosine ratio. *Pteridines* **2013**, *24*, 27–31. [CrossRef]
46. Barberis, E.; Timo, S.; Amede, E.; Vanella, V.V.; Puricelli, C.; Cappellano, G.; Raineri, D.; Cittone, M.G.; Rizzi, E.; Pedrinelli, A.R.; et al. Large-Scale plasma analysis revealed new mechanisms and molecules associated with the host response to SARS-CoV-2. *Int. J. Mol. Sci.* **2020**, *21*, 8623. [CrossRef]
47. Huang, C.; Huang, L.; Wang, Y.; Li, X.; Ren, L.; Gu, X.; Kang, L.; Guo, L.; Liu, M.; Zhou, X.; et al. 6-month consequences of COVID-19 in patients discharged from hospital: A cohort study. *Lancet* **2021**, *397*, 220–232. [CrossRef]

Article

Effects of Workers Exposure to Nanoparticles Studied by NMR Metabolomics

Štěpán Horník [1,2,†], Lenka Michálková [1,2,†], Jan Sýkora [1,*], Vladimír Ždímal [1], Štěpánka Vlčková [3], Štěpánka Dvořáčková [4] and Daniela Pelclová [3,*]

1 Institute of Chemical Process Fundamentals of the CAS, Rozvojová 1/135, 165 02 Prague, Czech Republic; hornik@icpf.cas.cz (Š.H.); michalkova@icpf.cas.cz (L.M.); zdimal@icpf.cas.cz (V.Ž.)
2 Department of Analytical Chemistry, University of Chemistry and Technology Prague, 166 28 Prague, Czech Republic
3 Department of Occupational Medicine, First Faculty of Medicine, Charles University in Prague and General University Hospital in Prague, Na Bojišti 1, 128 00 Prague, Czech Republic; stepanka.vlckova@vfn.cz
4 Department of Machining and Assembly, Department of Engineering Technology, Department of Material Science, Faculty of Mechanical Engineering, Technical University in Liberec, Studentska 1402/2, 461 17 Liberec, Czech Republic; stepanka.dvorackova@tul.cz
* Correspondence: sykora@icpf.cas.cz (J.S.); Daniela.Pelclova@lf1.cuni.cz (D.P.)
† These authors contributed equally to this study.

Abstract: In this study, the effects of occupational exposure to nanoparticles (NPs) were studied by NMR metabolomics. Exhaled breath condensate (EBC) and blood plasma samples were obtained from a research nanoparticles-processing unit at a national research university. The samples were taken from three groups of subjects: samples from workers exposed to nanoparticles collected before and after shift, and from controls not exposed to NPs. Altogether, 60 ^{1}H NMR spectra of exhaled breath condensate (EBC) samples and 60 ^{1}H NMR spectra of blood plasma samples were analysed, 20 in each group. The metabolites identified together with binning data were subjected to multivariate statistical analysis, which provided clear discrimination of the groups studied. Statistically significant metabolites responsible for group separation served as a foundation for analysis of impaired metabolic pathways. It was found that the acute effect of NPs exposure is mainly reflected in the pathways related to the production of antioxidants and other protective species, while the chronic effect is manifested mainly in the alteration of glutamine and glutamate metabolism, and the purine metabolism pathway.

Keywords: NMR metabolomics; human plasma; exhaled breath condensate; nanoparticles exposure

Citation: Horník, Š.; Michálková, L.; Sýkora, J.; Ždímal, V.; Vlčková, Š.; Dvořáčková, Š.; Pelclová, D. Effects of Workers Exposure to Nanoparticles Studied by NMR Metabolomics. *Appl. Sci.* **2021**, *11*, 6601. https://doi.org/10.3390/app11146601

Academic Editor: Alessia Vignoli

Received: 7 June 2021
Accepted: 14 July 2021
Published: 18 July 2021

1. Introduction

Nanoscience and nanotechnology have been developing rapidly in recent years, especially in new materials for electronics and optoelectronics fields, for energy technology, and in technology fields related to medical products, particularly for diagnostics and drugs delivery systems. The increased use of nanoparticles has raised concerns in many areas including the environment, human public health, consumer safety, and occupational safety and health [1,2]. Nanoparticles (NPs) are defined as particles with one or more dimensions at the nanoscale, less than 100 nm. The physiological response to NPs and the potential adverse effect on human health requires further research since contact with NPs is becoming a common part of everyday life. In recent years, numerous toxicity studies have assessed the hazard of NPs exposure [2–14]. In general, several health issues were associated with NPs including allergy, injury of epithelial tissue, inflammation, and oxidative stress response [1–3,6,10,11,15,16]. The mechanisms of NPs' biological interaction may vary according to the chemical composition, size, shape, bulk chemical composition, solubility, dose, etc. Moreover, NPs may show an increased toxicity when compared to larger particles of the same chemical composition that are little or even non-toxic by themselves [2,10,16,17].

For humans, inhalation is probably the most common way of NPs access followed by oral and dermal routes of exposure. Inhaled NPs can be deposited throughout the human respiratory system including pharyngeal, nasal, transbronchial and alveolar regions, depending on the particle size. The fractional deposition efficiency of particles below 100 nm is in the range of 30–70% in pulmonary regions, and the alveolar deposition increases as the size of NPs decreases [2,16–18]. After deposition in the respiratory tract, NPs may penetrate through membranes and thus enter the blood, pulmonary interstitium, brain, liver, heart, spleen and possibly to the foetus in pregnant females. Since NPs can have the same dimensions as some biomolecules, such as proteins and nucleic acids, adsorption and subsequent disruption of their structure are also possible [2,3,10,11,19].

The existing toxicological methodology for NPs still requires further adjustment to properly assess the risks, including the transport and distribution of NPs in the human body and the mechanism of interaction at the subcellular and molecular level, and to extrapolate the results from in vitro and animal models experiments, which may explain the human health deterioration. Another challenge of this field is to find a fast, specific and sensitive way to evaluate occupational risk. So far, the number of human studies is very limited. As the main exposure to NPs takes place via inhalation and the respiratory system is the primary afflicted organ system, collection and analysis of exhaled breath condensate (EBC) is the most frequently used non-invasive technique for assessment of a subject's condition. EBC contains, besides water, a small proportion of inorganic ions, small organic molecules, proteins and other macromolecules. Analysis of EBC enables the determination of important biomarkers as a response to current physiological conditions [20].

Recently, two toxicological studies were performed on a cohort of 20 workers exposed to NPs during their occupation [12,13]. A detailed analysis of lung function parameters obtained by spirometry revealed a significant decline of forced expiratory volume (FEV1) and its ratio to forced vital capacity (FVC) when compared to the pre-shift values or to the control group. These data were accompanied by LC-MS analysis of inflammation markers in EBC. The levels of pro-inflammatory markers LTB4, LTD4, LTE4, IL 9 and TNF were found to be increased in the worker group relative to controls. On the other hand, the levels of anti-inflammatory LXB4 and IL 10 were lower in the worker group than in controls. Moreover, the levels of the TNF (tumour necrosis factor) found in the pre-shift samples were positively correlated with the duration of employment in the NPs processing workshop [13]. LC-MS analysis was also targeted at markers of oxidative stress. The oxidation of lipids was evaluated from the levels of malondialdehyde (MDA), 4-hydroxy-*trans*-hexenal (HHE), 4-hydroxy-*trans*-nonenal (HNE), C6–C13 aldehydes, and 8-isoprostane; oxidative damage of nucleic acids from levels of 8-hydroxyguanosine (8-OHG), 8-hydroxy-2-deoxyguanosine (8-OHdG), and 5-hydroxymethyl uracil (5-OHMeU); oxidation of proteins from levels of *o*-tyrosine (*o*-Tyr), 3-chlorotyrosine (3-ClTyr), and 3-nitrotyrosine (3-NOTyr). A statistically significant increase was observed for all markers of lipid oxidation in post-shift samples relative to pre-shift ones, while the markers of oxidation of nucleic acids and proteins were found already significantly elevated in the pre-shift EBC samples, and no further increase was observed in the post-shift samples [12]. Both studies suggested lung impairment at the molecular level induced by oxidative stress associated with NPs exposure. However, the adverse effects were attributed rather to NPs in general than to specific chemical composition of NPs.

In this study, the EBC and blood plasma samples of the same cohort were analysed by ^{1}H NMR spectroscopy and processed by means of multivariate statistical analysis. It has already been shown that such NMR-based metabolomics can be advantageously used in NPs toxicology studies [1,11,21–28] reflecting the molecular changes induced by NPs inhalation. The samples studied here were examined as pre-shift and post-shift and were compared to controls. The main goal of this study was to assess the acute and chronic effect of NPs occupational exposure.

2. Materials and Methods

2.1. Workplace and Process Description

Subjects of the study were recruited at a research and development unit at a national research university, where a new thermoplastic or reactoplastic composite material was being developed. In the workplace, three different operations are performed, specifically, welding on metal surfaces, smelting of mixtures containing nanoadditives, and machining of the finished nanocomposite. A chemical analysis of aerosol generated in the working environment showed Fe, Mn and Si as the most abundant elements [12]. Aerosol mass concentration ranged from 0.12 to 1.84 mg/m^3 during nanocomposite machining processes. Median particle number concentration ranged from 4.8 to 105 $\times$ 10^6 particles/m^3 with the particle size ranging from 25 to 860 nm [12].

2.2. Subject Recruitment and Sample Collection

The samples were collected from 20 nanocomposite workers (15 men, 5 women; age 29–63, average 42 years; 1 smoker, 19 non-smokers) and from 20 control subjects living in the same district but working only in an office without any contact with NPs (13 men, 7 women; age 20–66, average 43 years; 2 smokers, 18 non-smokers).

The EBC and blood plasma samples from nanocomposite workers were collected twice during the workday, pre-shift (i.e., before 2.5 h exposure to NPs) and post-shift (i.e., after NPs exposure). The examinations are referred to as pre-shift and post-shift. Beside the NP exposed workplace, the rest of the 8-hour shift was spent in the office. The controls were examined only once during the same time frame as the workers.

The pre-shift samples were used to study the subacute/chronic effect on the subjects of exposures in previous days. Comparison of the pre-shift and post-shift samples was intended to evaluate the acute effect of exposure during the shift.

All subjects were asked questions from a standardized questionnaire which summarized information on personal and occupational history, medical treatments, dietary habits, smoking habits, and alcohol intake (Table S1, Supplementary Materials). Participants underwent a physical examination, followed by the collection of biological samples—exhaled breath condensate and blood plasma.

This study has been approved by the Ethics Committee of the 1st Medical Faculty, Charles University. All procedures were performed following the Helsinki Declaration and the Collection Law of the Czech Republic. All participants signed an informed consent.

2.3. EBC Collection

EBC samples were collected using an Ecoscreen Turbo DECCS device (Jaeger, Hochberg, Germany) equipped with a filter. All subjects breathed tidally for 15 min through a mouthpiece connected to the condenser (−20 °C) while wearing a nose-clip. A minimum volume of exhaled air of 120 L was monitored via the EcoVent device (Jaeger, Wurzburg, Germany). The sample collection took approximately 15 min. All samples were immediately frozen and stored at −80 °C.

2.4. Blood Plasma Collection

Venous blood (9 mL) from the subjects studied was collected using sterile blood collection tubes with heparin as an anticoagulant. The plasma fractions were obtained by centrifugation at 15,000$\times g$ for 10 min and immediately frozen and stored at −80 °C.

For more details on the subjects' cohort, working environment, analysis of NPs composition and properties, see previous publications [12,13]. A follow-up of the researchers in 2017 and 2018 confirmed the results from 2016 [29].

2.5. Sample Preparation

Samples were thawed at room temperature. For preparation of EBC and blood plasma samples for ^{1}H NMR analysis, the following operation procedures were determined.

2.6. EBC Sample Preparation

An aliquot of 500 μL of EBC was mixed with 100 μL phosphate buffer (0.1 mol/L^1, pH = 7.4, 0.1 mol/L^1 sodium salt of trimethylsilyl-2,2,3,3-d$_4$-propionic acid (TSP), 38 mmol/L^1 NaN_3). Thus, sufficient sample volume for NMR analysis was obtained and pH was adjusted to 7.7.

2.7. Blood Plasma Sample Preparation

Aliquots of 350 μL of blood plasma were centrifuged through an Amicon 3-kDa cut-off filter (Merck, Germany) for 30 min at 14,000 rpm to isolate low-molecular metabolites. Subsequently, the filtrate was mixed with 350 μL phosphate buffer in D_2O (0.1 mol/L^1, pH = 7.4, 0.1 mol/L^1 sodium salt of trimethylsilyl-2,2,3,3-d$_4$-propionic acid (TSP), 38 mmol/L^1 NaN_3). Thus, sufficient sample volume for NMR analysis was obtained and pH was adjusted to 7.4.

2.8. Acquisition

One dimensional proton NMR spectra for all EBC and plasma samples were acquired using a Varian INOVA 500 MHz spectrometer (Varian Instruments Inc., Palo Alto, CA, USA) operating at 499.87 MHz, equipped with Ultra Shim System II. A 5 mm probe with inner ^{1}H coil was used to maximize the sensitivity. Prior to the analysis, samples were kept for at least 10 min inside the NMR probe for temperature equilibration (298.15 K). The ^{1}H NMR spectra of EBC and plasma samples were obtained using wet1D and tnnoesy pulse sequence, respectively. Spectral width covered 8 kHz using 2.7 s acquisition time. A relaxation delay of 4 s and 2 s was used for EBC and plasma samples, respectively. The final spectrum resulted from an accumulation of 1000 scans. Representative ^{1}H NMR spectra can be found in Figures S1 and S2 in the Supplementary Materials.

2.9. Data Processing

The Fourier-transform spectra were manually corrected for phase and baseline distortions using Chenomx NMR Suite 8.0 (NMR Suite program, Edmonton, Alberta, Canada [30]). The experimental spectrum was referenced to TSP. The solvent signal residuum was subtracted, TSP signal linewidth was determined, and pH was set.

Compound profiling was performed in the Chenomx Profiler by precise fitting of the compounds from the Chenomx library to the experimental spectrum. In EBC samples, 15 metabolites were identified, while 58 metabolites were identified in blood plasma samples. Since only 15 metabolites were quantified in the EBC, binning was used for EBC spectra to obtain more variables per sample. The binning was applied to each spectrum in the range 0.7–8.6 ppm, except for the region containing residual water signal (4.1–5.6 ppm). Standard bin size of 0.02 ppm was used, yielding 320 bins.

The concentration data from plasma samples were normalized to the total concentration sum to reduce the effects of sample dilution prior to statistical analysis. Total area normalization works well in biofluids, in which overall concentrations of metabolites are almost constant among the samples, such as blood plasma or urine [31]. However, normalization to the total area is not recommended in the case of EBC samples because of large differences in dilution [32]. Hence, PQN normalization was used for the EBC samples as a more robust type of normalization [33].

2.10. Statistical Analyses

All data analyses were performed using the open-source software R [34] and Metaboanalyst 5.0 [35]. Multivariate data analyses were conducted on processed concentration data and binned data separately. As a first step of statistical analysis, principal component analysis (PCA) was used to provide preliminary insight on the data complexity, trends of grouping or identifying outliers. Subsequently, orthogonal partial least squares discriminant analysis (OPLS-DA) was used for sample classification. Multilevel partial least squares analysis (mPLS) was used in the case of comparison of the pre-shift and post-shift samples [36]. All reported values of accuracy, sensitivity, and specificity were assessed by means of 100 cycles of a Monte Carlo cross-validation scheme where 90% of the samples were randomly selected at each iteration as a training set to build the model; the remaining 10% were subsequently tested on performance characteristics for the classification.

In order to identify the most influential and statistically significant compounds, the Wilcoxon rank-sum test and its paired version, the Wilcoxon signed-rank test, were used. Obtained p-values were adjusted for multiple comparisons using the Benjamini and Hochberg correction [37]. The threshold of adjusted p-values was set to <0.05 for statistical significance. Fold change was performed following the general formula defined as a logarithm of base 2 of a division of a median concentration of an individual compound in one group by a median concentration of an individual compound in the other group. The result is projected in logarithm to base 2 scale.

Altered metabolic pathways were detected using Metaboanalyst 5.0 using the metabolite ID taken from the Human Metabolome Database. Metabolic pathway analysis was performed on blood plasma metabolic profiles to reveal the biological impact of NPs inhalation. A plot of affected pathways contained 43 nodes, each representing one pathway, with colour and size coding corresponding to pathway significance and its impact, respectively. The significance was generated from betweenness centrality and out-degree centrality measurements. The pathway impact was generated by the summation of importance measures of matched metabolites to all metabolites present within the pathway.

3. Results and Discussion

In this study, ^{1}H NMR spectra of 60 exhaled breath condensate (EBC) samples and 60 blood plasma samples were analysed. The samples originate from a research nanoparticles-processing unit at a national research university. The samples were taken from three groups of subjects: (i) samples from workers exposed to nanoparticles (NPs) collected before shift (pre-shift, 20 EBC and 20 blood plasma) and (ii) after shift (post-shift, 20 EBC and 20 blood plasma), and (iii) a control group of subjects not exposed to NPs (controls, 20 EBC and 20 blood plasma). The pre-shift and post-shift samples were collected from the same individuals. Individual groups are defined in Materials and Methods. A comparative study of the pre-shift and control samples was applied to reveal a subacute/chronic effect of NPs exposure, while the comparison of the pre-shift and post-shift samples should reflect the acute effect on the workers' health.

3.1. Exhaled Breath Condensate

Since exhaled breath condensate is composed of 99.9% water, the other constituents are rather diluted. For this reason, quantitative analysis of ^{1}H NMR spectra using the Chenomx reference library provided only 15 metabolites. Due to the limited number of metabolites quantified, a meaningful multivariate statistical analysis cannot be performed on such a dataset. However, univariate statistical analysis identified some of the metabolites as statistically significant for discrimination of the groups studied. A Wilcoxon rank-sum test showed that pre-shift and post-shift EBC samples are mainly characterized by significantly elevated levels of acetoin and propionate, and decreased acetone, isopropanol and lactate levels when compared to control samples (Table S2, Supplementary Materials). On the other hand, an increase in dimethylamine and decrease in acetoin are the most significant changes induced by NPs exposure as observed in comparison of pre-shift and post-shift EBC samples (Figure 1).

Final group discrimination analysis was performed using binning data. Fingerprinting of individual ^{1}H NMR spectra provided 320 bins which subsequently served as an input into multivariate statistical analysis. Principal component analysis (PCA) of all binned spectra did not show any significantly outlying sample. It also indicated certain trends in group separation; however, a clear discrimination was not achieved (Figure S3, Supplementary Materials). Satisfactory group separation was achieved by orthogonal partial least squares discriminant analysis (OPLS-DA), which was applied to the pre-shift and control group to reveal the chronic effect of NPs exposure and to the pre-shift/post-shift and post-shift/control group to uncover the acute effect.

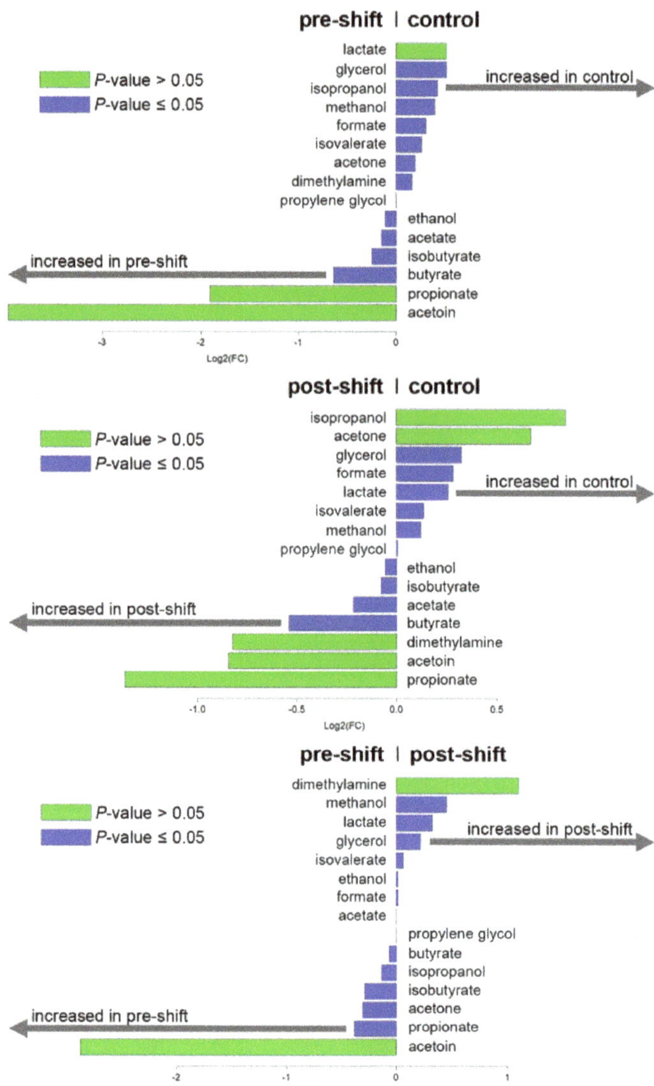

Figure 1. Fold change projections depicting differences in levels of individual metabolites observed in EBC samples between individual groups.

An excellent separation between the pre-shift and control group was achieved using three components. The model was characterized by 81.4% sensitivity, 94.8% specificity and 88.1% accuracy after Monte Carlo cross-validation (Figure 2a). Similarly, the separation of post-shift and the control group was achieved using a seven-component model yielding 88.7% accuracy, 93.9% sensitivity and 83.5% specificity after Monte Carlo cross-validation (Figure 2b).

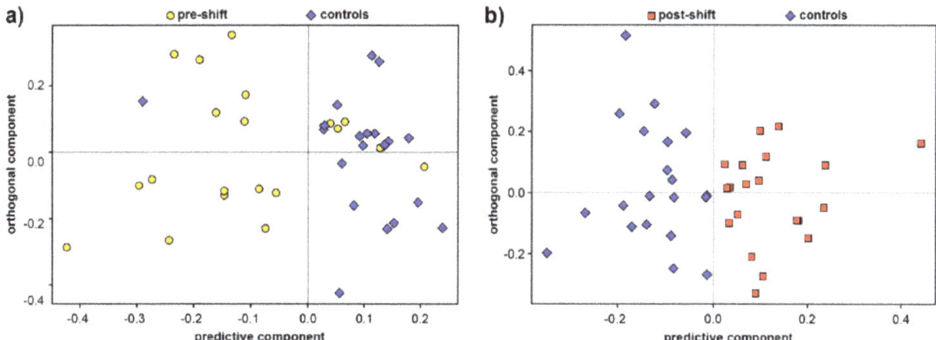

Figure 2. OPLS-DA of pre-shift subjects (yellow circles) and healthy controls (blue diamonds) using 320 bins from EBC samples; Acc. 88.1%, Sen. 81.4%, Spe. 94.8% (**a**). OPLS-DA of post-shift subjects (red squares) and healthy controls (blue diamonds) using 320 bins from EBC samples; Acc. 88.7%, Sen. 93.9%, Spe. 83.5% (**b**).

The bins contributing significantly to the group separation were identified from OPLS-DA loadings provided by Metaboanalyst. These bins show increased EBC concentration of acetoin, acetate and propionate in the pre-shift and post-shift samples when compared to the controls. Mainly increased signal intensities of alcohols were found in controls. These findings correspond well with the statistically significant compounds identified by univariate statistics as discussed above (Figure 1).

The comparison of the pre-shift and post-shift groups should reveal the acute effect of NPs exposure. The performed OPLS-DA provided a very good discrimination of the two groups using six components with accuracy of 83.1%, sensitivity of 84.1% and specificity of 82.1% after Monte Carlo cross-validation (Figure 3a). As both groups consist of the same 20 subjects whose samples were collected before and after the shift, a pairwise multilevel partial least squares (mPLS) analysis can be applied [36]. Compared to other PLS analyses, mPLS does not focus on investigation of the studied groups as a whole, but rather observes changes in each individual before and after the stimulus of the change and reflects the changes occurring within the same subject. The mPLS analysis showed a satisfactory discrimination of the two groups using three components with 82.0% accuracy after Monte Carlo cross-validation (Figure 3b). Although the OPLS and the mPLS models show similar accuracy, the mPLS model requires fewer components.

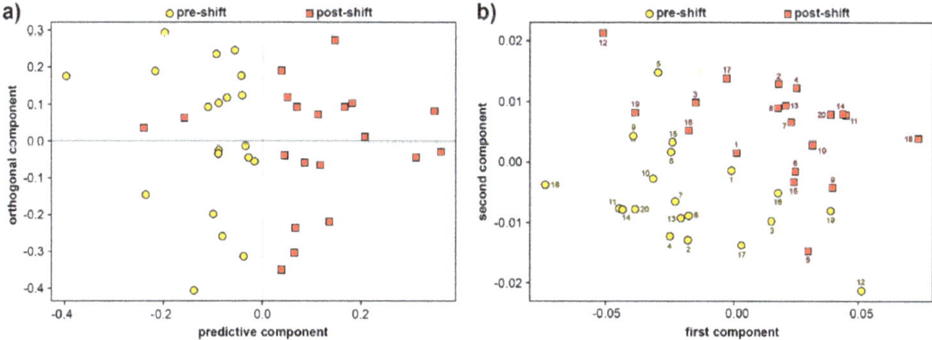

Figure 3. OPLS-DA of pre-shift (yellow circles) and post-shift subjects (red squares); Acc. 83.1%, Sen. 84.1%, Spe. 82.1% (**a**). Multilevel partial least squares (mPLS) analysis of pre-shift (yellow circles) and post-shift subjects (red squares); Acc. 82% (**b**). Both using 320 bins in each EBC sample.

The bins responsible for the group separation correspond to acetoin, which was found increased in the pre-shift group, and to lactate, formate and unsaturated chains of higher carboxylic acids increased in the post-shift group. This is in agreement with the statistically significant compounds identified by univariate statistics (Figure 1).

Acetoin is a commonly identified metabolite in EBC [38–40] as a product of the detoxification process of acetaldehyde [41].

Since dimethylamine was found increased only in the post-shift group, it is probable that it may be associated with the acute effect of NPs exposure.

The increased levels of short-chain fatty acids such as acetate, propionate and butyrate in NPs exposed groups in comparison to the control group could be attributed to involvement in the regulation of several leukocyte functions such as eicosanoids and cytokines/chemokines production [38]. Propionate is associated with lipid metabolism [39], which was also found affected by chronic exposure to NPs [12]. Boxplots of selected metabolites affected by NPs exposure are depicted in Figure 4.

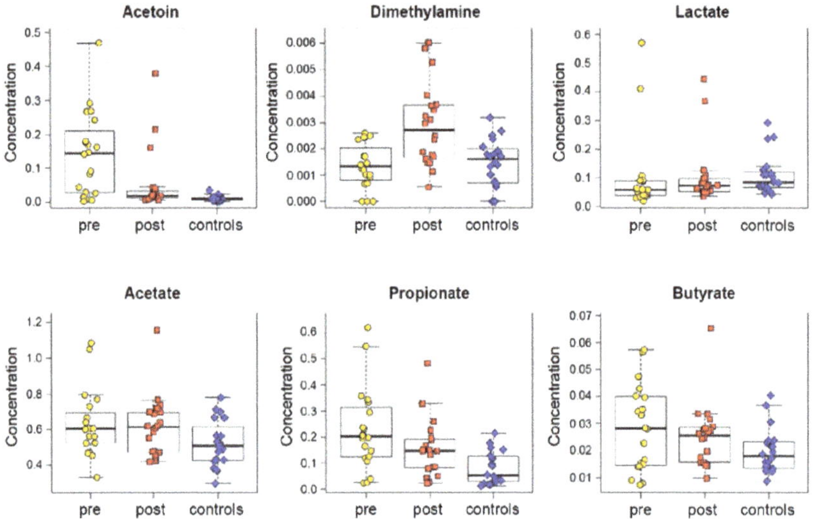

Figure 4. Boxplots of selected metabolites affected by NPs exposure.

3.2. Analysis of Blood Plasma

Using the Chenomx reference library, 58 metabolites were identified and quantified in each ^{1}H NMR spectrum of blood plasma samples. The concentration data of all quantified metabolites were used as an input for both multivariate and univariate statistical analyses to reveal important features of each group. The homogeneity of the groups was tested by principal component analysis (PCA) as an unsupervised statistical method. According to PCA, no sample was found significantly outlying. Nevertheless, group discrimination was not achieved (Figure S4, Supplementary Materials).

Subsequently, a supervised statistical method (OPLS-DA) was employed to pre-shift and control samples. A very good separation between these two groups was achieved using three components. The model was characterized by 88.2% sensitivity, 73.2% specificity and 80.7% accuracy after Monte Carlo cross-validation (Figure 5a).

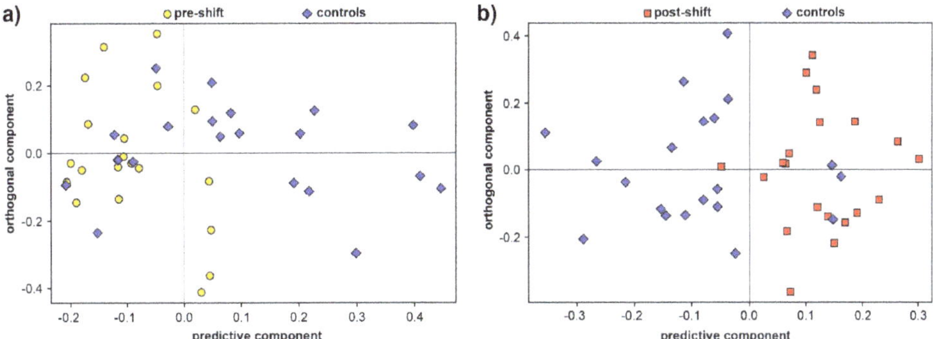

Figure 5. OPLS-DA of healthy controls (blue diamonds) and pre-shift subjects (yellow circles); Acc. 80.7%, Sen. 88.2%, Spe. 73.2% (**a**). OPLS-DA of post-shift subjects (red squares) and healthy controls (blue diamonds); Acc. 86.0%, Sen. 86.4%, Spe. 85.7% (**b**). Both using 58 normalized metabolites from blood plasma samples.

The nonparametric Wilcoxon rank-sum test was used to reveal statistically significant compounds that should reflect the effect of chronic exposure to NPs. Only acetone was found under the threshold for statistical significance (adjusted p-value ≤ 0.05). Four other metabolites were close to this threshold, specifically glutamate, glutamine, cystine and hypoxanthine (Table S3 and Figure S5 in Supplementary Materials). Levels of acetone, glutamine and cystine were found increased in the control group, whereas glutamate and hypoxanthine show higher levels in the pre-shift group.

OPLS-DA was also used for differentiation between the post-shift subjects and the healthy controls. A very good separation of the two groups was obtained using a six-component model with an accuracy of 86.0%, sensitivity of 86.4% and specificity of 85.7% after Monte Carlo cross-validation (Figure 5b). The nonparametric Wilcoxon rank-sum test was used to reveal statistically significant compounds (Table S3 and Figure S6 in Supplementary Materials). Seven metabolites were found under the threshold for statistical significance (adjusted p-value ≤ 0.05). Levels of propylene glycol, glutamate and pyruvate were found increased in the post-shift group, whereas acetone, mannose, 2-oxoisocaproate and *O*-acetylcarnitine showed higher levels in the control group.

Analogically, the acute effect of NPs exposure was also studied on plasma samples of the pre- and post-shift groups using OPLS-DA (Figure 6a). This discrimination analysis showed a certain potential to distinguish between the two groups with a model based on eight components characterized by 75.4% accuracy, 74.0% sensitivity and 76.9% specificity. Subsequently, mPLS was performed with a remarkable discrimination of the two groups using five components with 89.0% accuracy after Monte Carlo cross-validation (Figure 6b). In this case, mPLS reflects the intra-individual differences within each subject; therefore, it provides better group separation than discrimination analysis based on OPLS.

Subsequent analysis by the nonparametric pairwise Wilcoxon signed-rank test revealed 11 statistically significant compounds (Table S3 and Figure S7 in Supplementary Materials). Only compounds with adjusted p-value ≤ 0.05 after Benjamini-Hochberg correction were deemed statistically significant. Out of the 11 statistically significant compounds, increased levels for eight metabolites were found in the pre-shift group, specifically isobutyrate, 2-hydroxybutyrate, 2-oxoisocaproate, lactate, 3-hydroxybutyrate, isopropanol, tryptophan and 3-methyl-2-oxovalerate. Levels of three statistically significant metabolites were elevated in post-shift groups, namely propylene glycol, glycolate and myo-inositol.

The stress induced by NPs exposure is well documented by increased levels of certain metabolites in post-shift samples when compared to the pre-shift samples or controls. In particular, increased levels were found for propylene glycol, glycolate, myo-inositol, pyruvate and glutamate (Figure 7). Propylene glycol is known to be predominantly of exogenous origin as a part of various vitamins and other dietary supplements. The increased concentrations of propylene glycol in the post-shift group corresponds to the increased consumption of such supportive products in the morning before the shift.

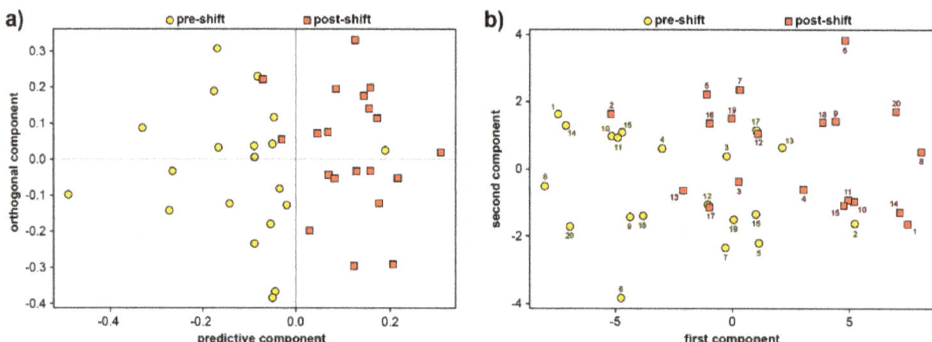

Figure 6. OPLS-DA of pre-shift (yellow circles) and post-shift subjects (red squares); Acc. 75.4%, Sen. 74.0%, Spe. 76.9% (**a**). mPLS analysis of pre-shift (yellow circles) and post-shift subjects (red squares); Acc. 89% (**b**). Both using 58 normalized metabolites from blood plasma samples.

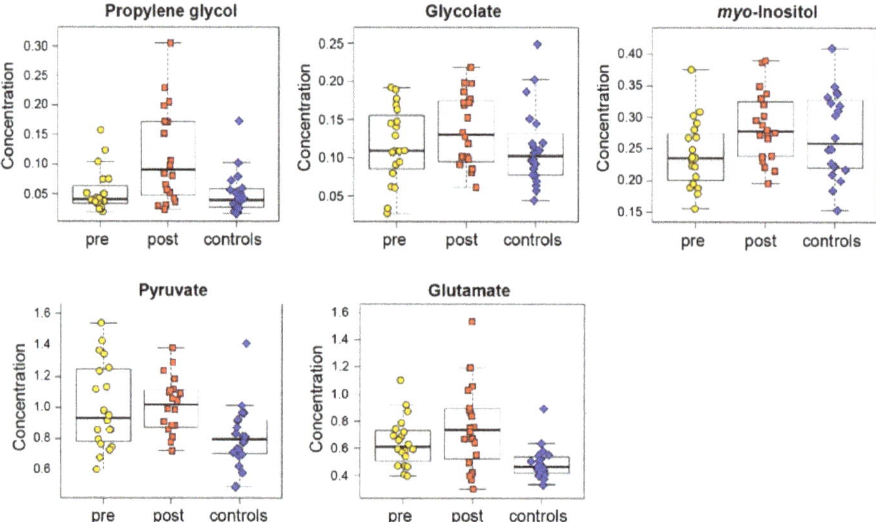

Figure 7. Boxplots of metabolites with increased levels in the post-shift samples.

On the other hand, the NPs exposure also induced a depletion of other metabolites in post-shift samples; namely of 3-methyl-2-oxovalerate, 2-oxoisocaproate, 2-hydroxybutyrate, 3-hydroxybutyrate, isobutyrate, isopropanol, mannose, *O*-acetylcarnitine and tryptophan (Figure 8). All the changes found in the post-shift group can be attributed to the acute effect of the NPs on workers' health.

The long-term effect of the NPs on workers' health can be deduced from the simultaneous changes in the pre- and post-shift group when compared to the control group. The levels of acetone, glutamine and cystine were found to decrease, while the levels of lactate and hypoxanthine increased in both groups when compared to the controls (Figure 9). The changes in levels of lactate and hypoxanthine were found to be more pronounced in the pre-shift group, which indicates involvement of these compounds in several metabolic pathways and mixing of acute and chronic effects.

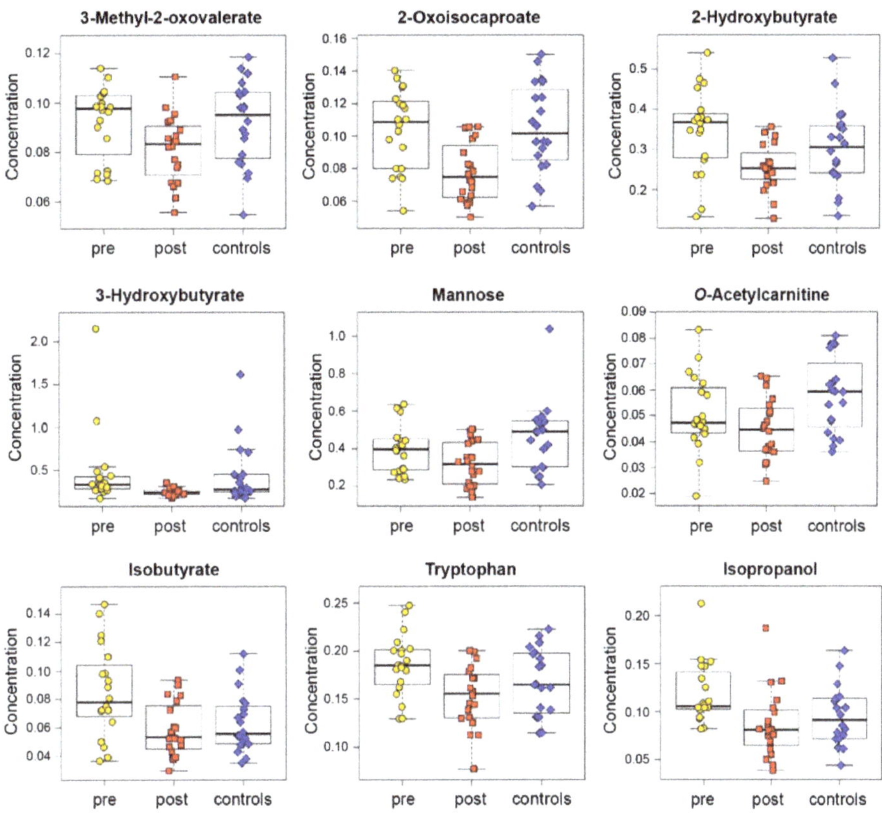

Figure 8. Boxplots of metabolites showing a depletion in the post-shift samples.

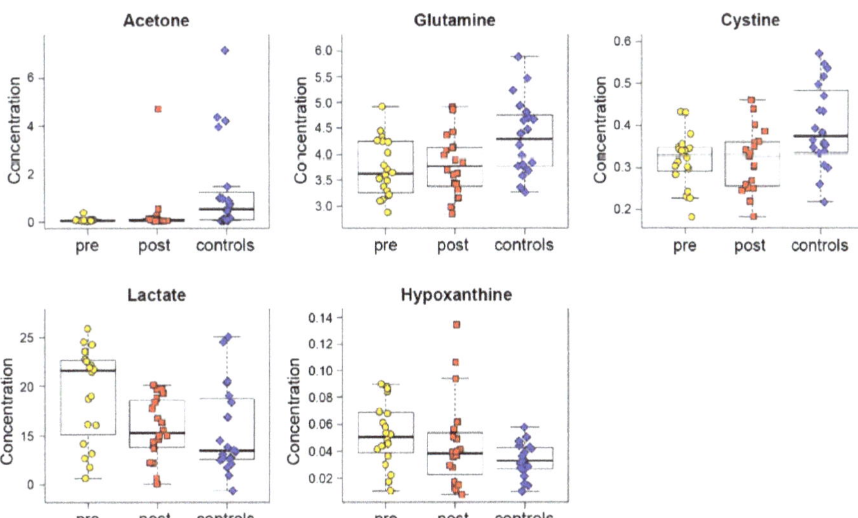

Figure 9. Boxplots of metabolites showing simultaneous changes in the pre- and post-shift group in comparison to the controls.

Metabolic pathway analysis was performed in MetaboAnalyst to assess involvement of the statistically significant metabolites in the individual metabolic pathways. The alterations between individual groups found by the pathway analysis are depicted in Figure S8 (Supplementary Materials) and the most affected pathways are summarized in Table S4 (Supplementary Materials).

The increased levels of lactate found in the pre-shift group can be partially associated with the metabolism of propylene glycol, which is contained in the food supplements administrated before the shift, as mentioned above. On the other hand, lactate is also involved in several other metabolic pathways, including pyruvate metabolism, according to pathway analysis (Figure S6 in Supplementary Materials) or glucose-alanine metabolism, according to the literature [42]. These pathways play an important role as energy pathways where lactate is usually produced from pyruvate. The highest concentrations of pyruvate were found in the post-shift group, indicating that the energy pathways were affected and the transformation between lactate and pyruvate is impacted by the NPs exposure. Increased levels of lactate have been observed in several studies on the impact of NPs exposure to rats [21,43,44]. Additionally, decreased levels of mannose, which can serve as an additional energy source, were observed in the post-shift group [45].

Elevated levels of lactate can lead to metabolic acidosis, similarly to increased levels of glycolate, which were found in the post-shift group. The formation of acid metabolites can induce inhibition of other metabolic pathways [46]. Glycolate is mainly involved in glyoxylate metabolism where it is oxidized to glyoxylate, which is further transformed into glycine [47]. Glyoxylate can be also transformed into oxalate, which is then caught and secreted by the renal tubules. Excessive concentrations of oxalate cause urolithiasis and nephrocalcinosis [48]. The excessive oxidation of glyoxylate to oxalate by lactate dehydrogenase is prevented by reduction of cytosolic and mitochondrial glyoxylate to glycolate by cytosolic glyoxylate reductase [48].

The post-shift group also manifested increased concentrations of myo-inositol. This metabolite has an important osmoregulatory role and is involved in the running of a wide range of cell functions, including cell growth and survival [49,50], which could explain why myo-inositol is increased in the acute state. Several studies have reported alterations in the myo-inositol levels after exposure to NPs in rats [43] or mouse fibroblast cells L929 [51].

The decreased tryptophan concentration in the post-shift group could be explained by its transformation to kynurenic acid via the kynurenine pathway (pathway tryptophan metabolism in Figure S8c in Supplementary Materials). Kynurenic acid has a protective effect against oxidative stress and lung inflammation induced by exposure to NPs [52]. Furthermore, the kynurenine pathway was previously associated with the elevated levels of cytokines [53], which is consistent with the results of our previous study [12,13]. A decreased concentration of tryptophan was also observed in rat blood serum after exposure to TiO_2 NPs [52], which is in agreement with the findings in workers exposed to nanoTiO_2, and similar oxidative stress effects [54,55].

The levels of several metabolites associated with the synthesis of glutathione were found altered, namely cystine, glutamate and 2-hydroxybutyrate. Glutathione as a major antioxidant is synthesized from cysteine, glutamate and glycine. Cysteine is transformed to glutathione in response to oxidative stress. This is reflected in decreased levels of cystine, an oxidized dimer form of cysteine. The availability of cysteine was reported as the rate-limiting step in the glutathione synthesis where cysteine is supplied via the cystine-glutamate antiporter system [49,56]. Elevated levels of glutamate in the pre- and post-shift group indicate that glutamate is exchanged for cystine in the antiporter system [49]. The elevated levels of glutamate also affect several other metabolic pathways, such as glutamine and glutamate metabolism, alanine, aspartate and glutamate metabolism, arginine and proline metabolism, histidine metabolism and butanoate metabolism, as is shown in the metabolic pathway analysis via MetaboAnalyst (Figure S8a,b in Supplementary Materials). Alterations in glutamate levels have also been observed in several studies focusing on NPs' impact on rats [26,43,51].

The utilization of glutamate in glutathione biosynthesis leads to higher demands on glutamate and increases glutamine transformation into glutamate via the glutamine and glutamate metabolism pathway. This is documented by decreased levels of glutamine in both pre- and post-shift groups compared to the control group. Together with glutamate, glutamine is also involved in the pathway of alanine, aspartate and glutamate metabolism, which was also found altered according to metabolic pathway analysis (Figure S8a,b in Supplementary Materials). Glutamine and glutamate were also found affected in the study of Kitchin et al. [57], in which the effect of TiO_2 and CeO_2 nanomaterials on human liver HepG2 cells was examined. A significant decrease in glutamine was observed, similar to observations from our study. Nevertheless, a decrease in glutamate was observed, on the contrary to our study, indicating that glutamate was involved at least partially in a different way.

2-Hydroxybutyrate is a reduction product of 2-ketobutyrate, which is produced during the transformation of cystathionine to cysteine within the methionine degradation pathway [58]. Since the concentration of 2-hydroxybutyrate is decreased in the acute state, 2-ketobutyrate is probably transformed into other metabolites, including propionyl-CoA [59], which is also associated with degradation of branched-chain amino acids, as discussed below.

The concentrations of two ketone bodies metabolites, 3-hydroxybutyrate and acetone, were found altered. Both compounds are closely connected to acetoacetate, another ketone body, which was, however, found unaltered. Nevertheless, decreased concentrations of 3-hydroxybutyrate and acetone after exposure to NPs suggest that the metabolic pathway of ketone body metabolism is affected. This was also revealed in the metabolic pathway analysis via MetaboAnalyst (Figure S8b in Supplementary Materials). 3-Hydroxybutyrate is also involved in butanoate metabolism, and it is also a degradation product of branched-chain amino acids, mainly of leucine [60]. The decreased concentrations of 3-hydroxybutyrate and acetone after NPs exposure are in contrast with other studies, which reported elevated levels of 3-hydroxybutyrate [21,23,61]. However, these studies were performed on rats exposed to high NPs doses. 3-Hydroxybutyrate is also an end product of β-oxidation of fatty acids [21]. The impairment of this metabolic pathway is also reflected in a decreased concentration of *O*-acetylcarnitine. This molecule serves as a carrier of acetyl from acetyl-CoA derived from fatty acids to mitochondria [62], thus taking part in energy metabolism. Moreover, decreased levels of *O*-acetylcarnitine can also be associated with oxidative stress. Similarly to this study, decreased levels of *O*-acetylcarnitine were also observed in zebrafish and mice embryos after Fe_2O_3 NPs exposure [62]. Accordingly, workers exposed to NPs during iron oxide pigment production showed elevated markers of lipid, nucleic acid, and protein oxidation in their EBC [63].

Decreased levels of 3-methyl-2-oxovalerate and 2-oxoisocaproate (4 methyl-2-oxovalerate) were found in the post-shift group. These compounds are produced as direct metabolites of isoleucine and leucine during their degradation by branched-chain amino acid aminotransferase [64]. Since the concentrations of leucine and isoleucine were found almost unaffected, leucine and isoleucine are involved in other pathways or processes, and the direct degradation pathway of these amino acids is inhibited. Isobutyrate is another metabolite associated with the metabolism of branched-chain amino acids, mainly of valine [65]. The decreased concentration of isobutyrate in the post-shift group also indicates that branched-chain amino acids' degradation is inhibited in the acute state. However, this inhibition was not manifested in the metabolic pathway analysis performed in MetaboAnalyst.

Hypoxanthine is an important part of purine metabolism [66], thus the elevated levels of hypoxanthine found mainly in pre-shift plasma samples indicate alterations in this metabolism. Such an observation is complementary to the findings of a previous study performed on the same cohort, where the markers of nucleic acids' oxidation (8-hydroxyguanosine and 8-hydroxy-2-deoxyguanosine) were identified in pre-shift EBC samples. Similar markers were also found in other toxicological studies of occupational exposure to different NPs, indicating a general effect of chronic NPs exposure [12,54,55].

The metabolic pathway analysis performed mainly indicates that the induced oxidative stress activates anti-oxidative pathways, and antioxidants, such as glutathione, are extensively consumed. Higher demands in the supplement of the consumed antioxidants can be observed in decreasing levels of their intermediates, in particular glutamine and cystine. The decreased tryptophan levels may be related to the production of its metabolites like kynurenic acid, which have protective effects against oxidative stress and lung inflammation. Moreover, alterations of several other metabolic pathways were observed.

The changes induced in metabolic profiles by NPs exposure were associated predominantly to the organism's response to oxidative stress. Similar response has been observed in studies dedicated to evaluation of the oxidative stress induced by smoking. Despite the number of smoking subjects being rather small in the presented study, statistical analysis was also performed after exclusion of the smoking subjects. One subject was excluded from the pre-shift/post-shift group and two subjects from the control group. The obtained results were in correspondence with those found in the original study and only minor changes were observed. Adjusted *p*-values of several metabolites levels previously found as statically significant raised slightly above the designated threshold. On the other hand, the adjusted *p*-value of hypoxanthine in blood plasma descended below the threshold in the comparison of the pre-shift and control groups (Tables S5 and S6 in Supplementary Materials). It is worth noting that the changes found in the results of univariate statistical analysis can be partially attributed to the decreased number of samples. The group separation provided by multivariate statistical analyses remained unaffected (Figures S9–S12 in Supplementary Materials). Major limitation of this study is the small number of subjects reflecting the actual size of the workplace, as all available workers were included.

4. Conclusions

The EBC and blood plasma samples of a cohort of 20 workers exposed to NPs during their occupation were analysed by ^{1}H NMR spectroscopy and processed by statistical analysis. Altogether, 15 metabolites were identified in EBC samples, while the analysis of plasma samples provided 58 metabolites. Subsequent multivariate statistical analyses performed on binning data from EBC and concentrations of 58 metabolites from plasma samples enabled clear discrimination between the pre-shift, post-shift and control groups. The univariate statistical analysis revealed statistically significant metabolites. Although plasma and EBC samples each showed changes in levels of different metabolites, the metabolic pathway analysis indicated, in both cases, mainly a reaction of the organism to oxidative stress and subsequent efforts for its protection.

The comparison of the pre-shift and post-shift group accompanied by comparison of the post-shift and control group provided insight into the acute effect of the NPs exposure. Altered levels of lactate, pyruvate, 3-hydroxybutyrate, mannose and *O*-acetylcarnithine indicated an energy balance impairment. The altered levels of glutamate, cystine, tryptophan, acetate, propionate and butyrate were associated to the pathways related to the production of antioxidants, mainly glutathione, and other protective species. The comparison of the pre-shift and control group revealed that the chronic effect of the NPs exposure manifested mainly in an alteration in glutamine and glutamate metabolism. The increased levels of hypoxanthine indicated an impairment of the purine metabolism pathway.

The presented results correspond well with similar studies performed on cohorts exposed to different types of NPs, indicating that the observed adverse effects can be attributed to nanoparticles in general, rather than to their chemical nature.

This work is one of the few dealing with the occupational exposure to NPs studied by the means of NMR metabolomics. Similar response to NPs exposure was observed for both types of samples indicating that either biofluid can be used for evaluation of adverse effects of nanoparticles inhalation. Potentially, blood derivatives could serve as an alternative to commonly used EBC samples.

Supplementary Materials: The following are available online at https://www.mdpi.com/article/10.3390/app11146601/s1, Table S1: Basic characteristics of the samples; Figure S1: 1H NMR spectrum of a representative EBC sample; Figure S2: 1H NMR spectrum of a representative blood plasma sample; Table S2: Wilcoxon test for EBC samples; Figure S3: Principal component analysis for EBC samples; Figure S4: Principal component analysis for plasma samples; Table S3: Wilcoxon test for blood plasma samples; Figure S5: Fold change projection of pre-shift subjects and healthy controls; Figure S6: Fold change projection of post-shift subjects and healthy controls; Figure S7: Fold change projection of pre-shift and post-shift subjects; Figure S8: Metabolic pathway analysis; Table S4: Overview of the most influenced metabolic pathways; Table S5: Wilcoxon test for EBC samples after exclusion of smoking subjects; Table S6: Wilcoxon test for blood plasma samples after exclusion of smoking subjects; Figures S9 and S10: Multivariate statistical analysis of EBC samples after exclusion of smoking subjects; Figures S11 and S12: Multivariate statistical analysis of plasma samples after exclusion of smoking subjects.

Author Contributions: Conceptualization, V.Ž., Š.D. and D.P.; methodology, Š.H., L.M. and J.S.; formal analysis, J.S. and D.P.; investigation, Š.H. and L.M.; resources, J.S., V.Ž. and D.P.; data curation, Š.H., L.M., J.S. and Š.V.; writing—original draft preparation, Š.H. and L.M.; writing—review and editing, J.S.; visualization, Š.H., L.M. and J.S.; supervision, J.S.; project administration, Š.V. and D.P.; funding acquisition, J.S., V.Ž. and D.P. All authors have read and agreed to the published version of the manuscript.

Funding: This work was supported by the Ministry of Youth, Education and Sports of the Czech Republic, project No. LM2018122 and by the Technology Agency of the Czech Republic (Grant No. TK02010035) and projects Progres Q25 and Q29 of the Charles University.

Institutional Review Board Statement: The study was conducted according to the guidelines of the Declaration of Helsinki, and approved by the Ethical Committee of 1st Medical Faculty, Charles University.

Informed Consent Statement: All participants were informed of the study aim at least five days earlier, and signed an informed consent form before the study began.

Data Availability Statement: The NMR spectra or evaluated data used in this study are available on request from the author: sykora@icpf.cas.cz.

Acknowledgments: The authors are grateful to Andrew Christensen for proofreading.

Conflicts of Interest: The authors declare no conflict of interest.

References

1. Schnackenberg, L.K.; Sun, J.; Beger, R.D. Metabolomics techniques in nanotoxicology studies. *Methods Mol. Biol.* **2012**, *926*, 141–156. [CrossRef]
2. Oberdörster, G.; Oberdörster, E.; Oberdörster, J. Nanotoxicology: An emerging discipline evolving from studies of ultrafine particles. *Environ. Health Perspect.* **2005**, *113*, 823–839. [CrossRef]
3. Oberdörster, G.; Maynard, A.; Donaldson, K.; Castranova, V.; Fitzpatrick, J.; Ausman, K.; Carter, J.; Karn, B.; Kreyling, W.; Lai, D.; et al. Principles for characterizing the potential human health effects from exposure to nanomaterials: Elements of a screening strategy. *Part. Fibre Toxicol.* **2005**, *2*, 8. [CrossRef]
4. Oberbek, P.; Kozikowski, P.; Czarnecka, K.; Sobiech, P.; Jakubiak, S.; Jankowski, T. Inhalation exposure to various nanoparticles in work environment—Contextual information and results of measurements. *J. Nanoparticle Res.* **2019**, *21*, 222. [CrossRef]
5. Geiser, M.; Jeannet, N.; Fierz, M.; Burtscher, H. Evaluating adverse effects of inhaled nanoparticles by realistic in vitro technology. *Nanomaterials* **2017**, *7*, 49. [CrossRef]
6. Kim, Y.J.; Yu, M.; Park, H.O.; Yang, S.I. Comparative study of cytotoxicity, oxidative stress and genotoxicity induced by silica nanomaterials in human neuronal cell line. *Mol. Cell. Toxicol.* **2010**, *6*, 337–344. [CrossRef]
7. Shvedova, A.; Castranova, V.; Kisin, E.; Murray, A.; Gandelsman, V.; Baron, P. Journal of Toxicology and Environmental Health, Part A: Current Issues Exposure to Carbon Nanotube Material: Assessment of Nanotube Cytotoxicity using Human Keratinocyte Cells. *J. Toxocol. Environ. Health Part A* **2011**, *66*, 1909–1926. [CrossRef]
8. Brown, D.M.; Donaldson, K.; Borm, P.J.; Schins, R.P.; Dehnhardt, M.; Gilmour, P.; Jimenez, L.A.; Stone, V. Calcium and ROS-mediated activation of transcription factors and TNF-α cytokine gene expression in macrophages exposed to ultrafine particles. *Am. J. Physiol. Lung Cell. Mol. Physiol.* **2004**, *286*, 344–353. [CrossRef] [PubMed]
9. Dick, C.A.J.; Brown, D.M.; Donaldson, K.; Stone, V. The Role of Free Radicals in the Toxic and Inflammatory Effects of Four Different Ultrafine Particle Types. *Inhal. Toxicol.* **2003**, *15*, 39–52. [CrossRef] [PubMed]
10. Nel, A.; Xia, T.; Mädler, L.; Li, N. Toxic Potential of Materials at the Nanolevel. *Science* **2006**, *311*, 622–627. [CrossRef] [PubMed]

11. Fröhlich, E. Role of omics techniques in the toxicity testing of nanoparticles. *J. Nanobiotechnol.* **2017**, *15*, 84. [CrossRef]
12. Pelclova, D.; Zdimal, V.; Schwarz, J.; Dvorackova, S.; Komarc, M.; Ondracek, J.; Kostejn, M.; Kacer, P.; Vlckova, S.; Fenclova, Z.; et al. Markers of oxidative stress in the exhaled breath condensate of workers handling nanocomposites. *Nanomaterials* **2018**, *8*, 611. [CrossRef] [PubMed]
13. Pelclova, D.; Zdimal, V.; Komarc, M.; Vlckova, S.; Fenclova, Z.; Ondracek, J.; Schwarz, J.; Kostejn, M.; Kacer, P.; Dvorackova, S.; et al. Deep airway inflammation and respiratory disorders in nanocomposite workers. *Nanomaterials* **2018**, *8*, 731. [CrossRef]
14. Fröhlich, E. Comparison of conventional and advanced in vitro models in the toxicity testing of nanoparticles. *Artif. Cells Nanomed. Biotechnol.* **2018**, *46*, 1091–1107. [CrossRef]
15. Dybing, E.; Løvdal, T.; Hetland, R.B.; Løvik, M.; Schwarze, P.E. Respiratory allergy adjuvant and inflammatory effects of urban ambient particles. *Toxicology* **2004**, *198*, 307–314. [CrossRef]
16. Seaton, A.; Tran, L.; Aitken, R.; Donaldson, K. Nanoparticles, human health hazard and regulation. *J. R. Soc. Interface* **2010**, *7*, S119–S129. [CrossRef]
17. Oberdörster, G.; Kuhlbusch, T.A.J. In vivo effects: Methodologies and biokinetics of inhaled nanomaterials. *NanoImpact* **2018**, *10*, 38–60. [CrossRef]
18. Cassee, F.R.; Muijser, H.; Duistermaat, E.; Freijer, J.J.; Geerse, K.B.; Marijnissen, J.C.; Arts, J.H. Particle size-dependent total mass deposition in lungs determines inhalation toxicity of cadmium chloride aerosols in rats. Application of a multiple path dosimetry model. *Arch. Toxicol.* **2002**, *76*, 277–286. [CrossRef] [PubMed]
19. Buckley, A.; Hodgson, A.; Warren, J.; Guo, C.; Smith, R. Size-dependent deposition of inhaled nanoparticles in the rat respiratory tract using a new nose-only exposure system. *Aerosol Sci. Technol.* **2016**, *50*, 1–10. [CrossRef]
20. Horváth, I.; Barnes, P.J.; Loukides, S.; Sterk, P.J.; Högman, M.; Olin, A.C.; Amann, A.; Antus, B.; Baraldi, E.; Bikov, A.; et al. A European respiratory society technical standard: Exhaled biomarkers in lung disease. *Eur. Respir. J.* **2017**, *49*, 1600965. [CrossRef]
21. Lee, S.H.; Wang, T.Y.; Hong, J.H.; Cheng, T.J.; Lin, C.Y. NMR-based metabolomics to determine acute inhalation effects of nano- and fine-sized ZnO particles in the rat lung. *Nanotoxicology* **2016**, *10*, 924–934. [CrossRef]
22. Li, X.; Ban, Z.; Yu, F.; Hao, W.; Hu, X. Untargeted Metabolic Pathway Analysis as an Effective Strategy to Connect Various Nanoparticle Properties to Nanoparticle-Induced Ecotoxicity. *Environ. Sci. Technol.* **2020**, *54*, 3395–3406. [CrossRef]
23. Li, J.; Zhao, Z.; Feng, J.; Gao, J.; Chen, Z. Understanding the metabolic fate and assessing the biosafety of MnO nanoparticles by metabonomic analysis. *Nanotechnology* **2013**, *24*, 455102. [CrossRef]
24. Zhang, W.; Zhao, Y.; Li, F.; Li, L.; Feng, Y.; Min, L.; Ma, D.; Yu, S.; Liu, J.; Zhang, H.; et al. Zinc oxide nanoparticle caused plasma metabolomic perturbations correlate with hepatic steatosis. *Front. Pharmacol.* **2018**, *9*, 57. [CrossRef]
25. Dailey, L.A.; Hernández-Prieto, R.; Casas-Ferreira, A.M.; Jones, M.-C.; Riffo-Vasquez, Y.; Rodríguez-Gonzalo, E.; Spina, D.; Jones, S.A.; Smith, N.W.; Forbes, B.; et al. Adenosine monophosphate is elevated in the bronchoalveolar lavage fluid of mice with acute respiratory toxicity induced by nanoparticles with high surface hydrophobicity. *Nanotoxicology* **2015**, *9*, 106–115. [CrossRef]
26. Guo, Z.; Luo, Y.; Zhang, P.; Chetwynd, A.J.; Qunhui Xie, H.; Monikh, F.A.; Tao, W.; Xie, C.; Liu, Y.; Xu, L.; et al. Deciphering the particle specific effects on metabolism in rat liver and plasma from ZnO nanoparticles versus ionic Zn exposure. *Environ. Int.* **2020**, *136*, 105437. [CrossRef] [PubMed]
27. Cui, L.; Wang, X.; Sun, B.; Xia, T.; Hu, S. Predictive Metabolomic Signatures for Safety Assessment of Metal Oxide Nanoparticles. *ACS Nano* **2019**, *13*, 13065–13082. [CrossRef] [PubMed]
28. Guo, N.L.; Poh, T.Y.; Pirela, S.; Farcas, M.T.; Chotirmall, S.H.; Tham, W.K.; Adav, S.S.; Ye, Q.; Wei, Y.; Shen, S.; et al. Integrated transcriptomics, metabolomics, and lipidomics profiling in rat lung, blood, and serum for assessment of laser printer-emitted nanoparticle inhalation exposure-induced disease risks. *Int. J. Mol. Sci.* **2019**, *20*, 6384. [CrossRef] [PubMed]
29. Pelclova, D.; Zdimal, V.; Komarc, M.; Schwarz, J.; Ondracek, J.; Ondrackova, L.; Kostejn, M.; Vlckova, S.; Fenclova, Z.; Dvorackova, S.; et al. Three-year study of markers of oxidative stress in exhaled breath condensate in workers producing nanocomposites, extended by plasma and urine analysis in last two years. *Nanomaterials* **2020**, *10*, 2440. [CrossRef]
30. Chenomx Inc. *ChenomX NMR Suite 8.0*; Chenomx Inc.: Edmonton, AB, Canada, 2016.
31. Bertini, I.; Luchinat, C.; Miniati, M.; Monti, S.; Tenori, L. Phenotyping COPD by 1H NMR metabolomics of exhaled breath condensate. *Metabolomics* **2014**, *10*, 302–311. [CrossRef]
32. Vignoli, A.; Ghini, V.; Meoni, G.; Licari, C.; Takis, P.G.; Tenori, L.; Turano, P.; Luchinat, C. High-Throughput Metabolomics by 1D NMR. *Angew. Chem. Int. Ed.* **2019**, *58*, 968–994. [CrossRef]
33. Dieterle, F.; Ross, A.; Schlotterbeck, G.; Senn, H. Probabilistic Quotient Normalization as Robust Method to Account for Dilution of Complex Biological Mixtures. Application in 1H NMR Metabonomics. *Anal. Chem.* **2006**, *78*, 4281–4290. [CrossRef]
34. RC Team. *R: A Language and Environment for Statistical Computing*; R Foundation for Statistical Computing: Vienna, Austria, 2017.
35. Pang, Z.; Chong, J.; Zhou, G.; de Lima Morais, D.A.; Chang, L.; Barrette, M.; Gauthier, C.; Jacques, P.-É.; Li, S.; Xia, J. MetaboAnalyst 5.0: Narrowing the gap between raw spectra and functional insights. *Nucleic Acids Res.* **2021**, *49*, W388–W396. [CrossRef]
36. Westerhuis, J.A.; van Velzen, E.J.J.; Hoefsloot, H.C.J.; Smilde, A.K. Multivariate paired data analysis: Multilevel PLSDA versus OPLSDA. *Metabolomics* **2010**, *6*, 119–128. [CrossRef]
37. Benjamini, Y.; Hochberg, Y. Controlling the False Discovery Rate: A Practical and Powerful Approach to Multiple Testing. *J. R. Stat. Soc. Ser. B* **1995**, *57*, 289–300. [CrossRef]
38. de Laurentiis, G.; Paris, D.; Melck, D.; Montuschi, P.; Maniscalco, M.; Bianco, A.; Sofia, M.; Motta, A. Separating Smoking-Related Diseases Using NMR-Based Metabolomics of Exhaled Breath Condensate. *J. Proteome Res.* **2013**, *12*, 1502–1511. [CrossRef]

39. Airoldi, C.; Ciaramelli, C.; Fumagalli, M.; Bussei, R.; Mazzoni, V.; Viglio, S.; Iadarola, P.; Stolk, J. 1H NMR to Explore the Metabolome of Exhaled Breath Condensate in α1-Antitrypsin Deficient Patients: A Pilot Study. *J. Proteome Res.* **2016**, *15*, 4569–4578. [CrossRef]
40. Vignoli, A.; Santini, G.; Tenori, L.; Macis, G.; Mores, N.; Macagno, F.; Pagano, F.; Higenbottam, T.; Luchinat, C.; Montuschi, P. NMR-Based Metabolomics for the Assessment of Inhaled Pharmacotherapy in Chronic Obstructive Pulmonary Disease Patients. *J. Proteome Res.* **2020**, *19*, 64–74. [CrossRef] [PubMed]
41. Otsuka, M.; Harada, N.; Itabashi, T.; Ohmori, S. Blood and urinary levels of ethanol, acetaldehyde, and C4 compounds such as diacetyl, acetoin, and 2,3-butanediol in normal male students after ethanol ingestion. *Alcohol* **1999**, *17*, 119–124. [CrossRef]
42. Sun, M.; Zhang, J.; Liang, S.; Du, Z.; Liu, J.; Sun, Z.; Duan, J. Metabolomic characteristics of hepatotoxicity in rats induced by silica nanoparticles. *Ecotoxicol. Environ. Saf.* **2021**, *208*, 111496. [CrossRef] [PubMed]
43. Feng, J.; Liu, H.; Bhakoo, K.K.; Lu, L.; Chen, Z. A metabonomic analysis of organ specific response to USPIO administration. *Biomaterials* **2011**, *32*, 6558–6569. [CrossRef] [PubMed]
44. Lin, B.; Zhang, H.; Lin, Z.; Fang, Y.; Tian, L.; Yang, H.; Yan, J.; Liu, H.; Zhang, W.; Xi, Z. Studies of single-walled carbon nanotubes-induced hepatotoxicity by NMR-based metabonomics of rat blood plasma and liver extracts. *Nanoscale Res. Lett.* **2013**, *8*, 236. [CrossRef] [PubMed]
45. Sharma, V.; Ichikawa, M.; Freeze, H.H. Mannose metabolism: More than meets the eye. *Biochem. Biophys. Res. Commun.* **2014**, *453*, 220–228. [CrossRef]
46. Brent, J. Current Management of Ethylene Glycol Poisoning. *Drugs* **2001**, *61*, 979–988. [CrossRef]
47. Meléndez-Hevia, E.; De Paz-Lugo, P.; Cornish-Bowden, A.; Cárdenas, M.L. A weak link in metabolism: The metabolic capacity for glycine biosynthesis does not satisfy the need for collagen synthesis. *J. Biosci.* **2009**, *34*, 853–872. [CrossRef] [PubMed]
48. Salido, E.; Pey, A.L.; Rodriguez, R.; Lorenzo, V. Primary hyperoxalurias: Disorders of glyoxylate detoxification. *Biochim. Biophys. Acta Mol. Basis Dis.* **2012**, *1822*, 1453–1464. [CrossRef]
49. Carrola, J.; Bastos, V.; Daniel-da-Silva, A.L.; Gil, A.M.; Santos, C.; Oliveira, H.; Duarte, I.F. Macrophage Metabolomics Reveals Differential Metabolic Responses to Subtoxic Levels of Silver Nanoparticles and Ionic Silver. *Eur. J. Inorg. Chem.* **2020**, *2020*, 1867–1876. [CrossRef]
50. Chhetri, D.R. Myo-inositol and its derivatives: Their emerging role in the treatment of human diseases. *Front. Pharmacol.* **2019**, *10*, 1172. [CrossRef]
51. Bo, Y.; Jin, C.; Liu, Y.; Yu, W.; Kang, H. Metabolomic analysis on the toxicological effects of TiO_2 nanoparticles in mouse fibroblast cells: From the perspective of perturbations in amino acid metabolism. *Toxicol. Mech. Methods* **2014**, *24*, 461–469. [CrossRef]
52. Chen, Z.; Han, S.; Zhou, D.; Zheng, P.; Zhou, S.; Jia, G. Serum metabolomic signatures of Sprague-Dawley rats after oral administration of titanium dioxide nanoparticles. *NanoImpact* **2020**, *19*, 100236. [CrossRef]
53. Pedraz-Petrozzi, B.; Elyamany, O.; Rummel, C.; Mulert, C. Effects of inflammation on the kynurenine pathway in schizophrenia—A systematic review. *J. Neuroinflammation* **2020**, *17*, 56. [CrossRef] [PubMed]
54. Pelclova, D.; Zdimal, V.; Fenclova, Z.; Vlckova, S.; Turci, F.; Corazzari, I.; Kacer, P.; Schwarz, J.; Zikova, N.; Makes, O.; et al. Markers of oxidative damage of nucleic acids and proteins among workers exposed to TiO_2 (nano) particles. *Occup. Environ. Med.* **2016**, *73*, 110–118. [CrossRef] [PubMed]
55. Pelclova, D.; Zdimal, V.; Kacer, P.; Zikova, N.; Komarc, M.; Fenclova, Z.; Vlckova, S.; Schwarz, J.; Makeš, O.; Syslova, K.; et al. Markers of lipid oxidative damage in the exhaled breath condensate of nano TiO_2 production workers. *Nanotoxicology* **2017**, *11*, 52–63. [CrossRef] [PubMed]
56. Klomsiri, C.; Karplus, P.A.; Poole, L.B. Cysteine-based redox switches in enzymes. *Antioxid. Redox Signal.* **2011**, *14*, 1065–1077. [CrossRef] [PubMed]
57. Kitchin, K.T.; Grulke, E.; Robinette, B.L.; Castellon, B.T. Metabolomic effects in HepG2 cells exposed to four TiO_2 and two CeO_2 nanomaterials. *Environ. Sci. Nano* **2014**, *1*, 466–477. [CrossRef]
58. Grapov, D.; Fiehn, O.; Campbell, C.; Chandler, C.J.; Burnett, D.J.; Souza, E.C.; Casazza, G.A.; Keim, N.L.; Newman, J.W.; Hunter, G.R.; et al. Exercise plasma metabolomics and xenometabolomics in obese, sedentary, insulin-resistant women: Impact of a fitness and weight loss intervention. *Am. J. Physiol. Endocrinol. Metab.* **2019**, *317*, E999–E1014. [CrossRef]
59. Adams, S.H. Emerging perspectives on essential amino acid metabolism in obesity and the insulin-resistant state. *Adv. Nutr.* **2011**, *2*, 445–456. [CrossRef]
60. Worrall, E.B.; Gassain, S.; Cox, D.J.; Sugden, M.C.; Palmer, T.N. 3-Hydroxyisobutyrate dehydrogenase, an impurity in commercial 3-hydroxybutyrate dehydrogenase. *Biochem. J.* **1987**, *241*, 297–300. [CrossRef]
61. Bu, Q.; Yan, G.; Deng, P.; Peng, F.; Lin, H.; Xu, Y.; Cao, Z.; Zhou, T.; Xue, A.; Wang, Y.; et al. NMR-based metabonomic study of the sub-acute toxicity of titanium dioxide nanoparticles in rats after oral administration. *Nanotechnology* **2010**, *21*, 125105. [CrossRef]
62. Huang, Z.; Xu, B.; Huang, X.; Zhang, Y.; Yu, M.; Han, X.; Song, L.; Xia, Y.; Zhou, Z.; Wang, X.; et al. Metabolomics reveals the role of acetyl-l-carnitine metabolism in γ-Fe2O3 NP-induced embryonic development toxicity via mitochondria damage. *Nanotoxicology* **2019**, *13*, 204–220. [CrossRef] [PubMed]
63. Pelclova, D.; Zdimal, V.; Kacer, P.; Fenclova, Z.; Vlckova, S.; Syslova, K.; Navratil, T.; Schwarz, J.; Zikova, N.; Barosova, H.; et al. Oxidative stress markers are elevated in exhaled breath condensate of workers exposed to nanoparticles during iron oxide pigment production. *J. Breath Res.* **2016**, *10*, 016004. [CrossRef] [PubMed]

64. Islam, M.M.; Wallin, R.; Wynn, R.M.; Conway, M.; Fujii, H.; Mobley, J.A.; Chuang, D.T.; Hutson, S.M. A novel branched-chain amino acid metabolon: Protein-protein interactions in a supramolecular complex. *J. Biol. Chem.* **2007**, *282*, 11893–11903. [CrossRef] [PubMed]
65. Oliphant, K.; Allen-Vercoe, E. Macronutrient metabolism by the human gut microbiome: Major fermentation by-products and their impact on host health. *Microbiome* **2019**, *7*, 91. [CrossRef] [PubMed]
66. Pang, B.; McFaline, J.L.; Burgis, N.E.; Dong, M.; Taghizadeh, K.; Sullivan, M.R.; Elmquist, C.E.; Cunningham, R.P.; Dedon, P.C. Defects in purine nucleotide metabolism lead to substantial incorporation of xanthine and hypoxanthine into DNA and RNA. *Proc. Natl. Acad. Sci. USA* **2012**, *109*, 2319–2324. [CrossRef] [PubMed]

 applied sciences

Article

Exploring Serum NMR-Based Metabolomic Fingerprint of Colorectal Cancer Patients: Effects of Surgery and Possible Associations with Cancer Relapse

Alessia Vignoli [1,2,†], Elena Mori [3,†], Samantha Di Donato [3], Luca Malorni [3,4], Chiara Biagioni [5], Matteo Benelli [5], Vanessa Calamai [3], Stefano Cantafio [6], Annamaria Parnofiello [3,7], Maddalena Baraghini [6], Alessia Garzi [6], Francesca Del Monte [3], Dario Romagnoli [5], Ilenia Migliaccio [4], Claudio Luchinat [1,2], Leonardo Tenori [1,2,*] and Laura Biganzoli [3,*]

1 Magnetic Resonance Center (CERM) and Department of Chemistry "Ugo Schiff", University of Florence, 50019 Sesto Fiorentino, Italy; vignoli@cerm.unifi.it (A.V.); luchinat@cerm.unifi.it (C.L.)
2 Consorzio Interuniversitario Risonanze Magnetiche Metallo Proteine (CIRMMP), 50019 Sesto Fiorentino, Italy
3 Department of Medical Oncology, New Hospital of Prato-S. Stefano, 59100 Prato, Italy; elena2.mori@uslcentro.toscana.it (E.M.); samantha.didonato@uslcentro.toscana.it (S.D.D.); luca.malorni@uslcentro.toscana.it (L.M.); vanessa.calamai@uslcentro.toscana.it (V.C.); annamaria.parnofiello@uslcentro.toscana.it (A.P.); francesca.delmonte@uslcentro.toscana.it (F.D.M.)
4 "Sandro Pitigliani" Translational Research Unit, New Hospital of Prato-S. Stefano, 59100 Prato, Italy; ilenia.migliaccio@uslcentro.toscana.it
5 Bioinformatics Unit, Medical Oncology Department, New Hospital of Prato-S. Stefano, 59100 Prato, Italy; chiara.biagioni@uslcentro.toscana.it (C.B.); matteo.benelli@uslcentro.toscana.it (M.B.); dario.romagnoli@uslcentro.toscana.it (D.R.)
6 Department of Surgery, New Hospital of Prato-S. Stefano, 59100 Prato, Italy; stefano.cantafio@uslcentro.toscana.it (S.C.); maddalena.baraghini@uslcentro.toscana.it (M.B.); alessia.garzi@uslcentro.toscana.it (A.G.)
7 Department of Medicine (DAME), University of Udine, 33100 Udine, Italy
* Correspondence: tenori@cerm.unifi.it (L.T.); laura.biganzoli@uslcentro.toscana.it (L.B.); Tel.: +39-0554574281 (L.T.); +39-0574802520 (L.B.)
† Co-first authors.

Citation: Vignoli, A.; Mori, E.; Di Donato, S.; Malorni, L.; Biagioni, C.; Benelli, M.; Calamai, V.; Cantafio, S.; Parnofiello, A.; Baraghini, M.; et al. Exploring Serum NMR-Based Metabolomic Fingerprint of Colorectal Cancer Patients: Effects of Surgery and Possible Associations with Cancer Relapse. *Appl. Sci.* **2021**, *11*, 11120. https://doi.org/10.3390/app112311120

Academic Editor: John Patrick Alao

Received: 21 October 2021
Accepted: 18 November 2021
Published: 23 November 2021

Abstract: Background: Colorectal cancer (CRC) is the fourth most commonly diagnosed and third most deadly cancer worldwide. Surgery is the main treatment option for early disease; however, a relevant proportion of CRC patients relapse. Here, variations among preoperative and postoperative serum metabolomic fingerprint of CRC patients were studied, and possible associations between metabolic variations and cancer relapse were explored. Methods: A total of 41 patients with stage I–III CRC, planned for radical resection, were enrolled. Serum samples, collected preoperatively (t0) and 4–6 weeks after surgery before the start of any treatment (t1), were analyzed via NMR spectroscopy. NMR data were analyzed using multivariate and univariate statistical approaches. Results: Serum metabolomic fingerprints show differential clustering between t0 and t1 (82–85% accuracy). Pyruvate, HDL-related parameters, acetone, and 3-hydroxybutyrate appear to be the major players in this discrimination. Eight out of the 41 CRC patients enrolled developed cancer relapse. Postoperative, relapsed patients show an increase of pyruvate and HDL-related parameters, and a decrease of Apo-A1 Apo-B100 ratio and VLDL-related parameters. Conclusions: Surgery significantly alters the metabolomic fingerprint of CRC patients. Some metabolic changes seem to be associated with the development of cancer relapse. These data, if validated in a larger cohort, open new possibilities for risk stratification in patients with early-stage CRC.

Keywords: metabolomics; colorectal cancer; nuclear magnetic resonance; surgery; relapse

1. Introduction

Colorectal cancer (CRC) is the third most frequently diagnosed cancer and the second leading cause of cancer death worldwide [1–3]. A total of 80% of colon cancers are diagnosed at early stage (stage 1 to 3), and surgery is the primary treatment option with curative intent for this type of disease [4]. Unfortunately, about 35% of these patients develop cancer relapse, which, in the majority of cases, occurs within the first 2–3 years after surgery [5,6]. TNM staging at diagnosis, based on depth of tumor wall invasion (T), lymph node involvement (N), and presence of distant metastasis (M), is currently the principal instrument available to predict risk of relapse, and thus to identify patients who may have potential benefits from adjuvant treatment [7,8].

Colorectal cancer is a heterogeneous disease, even within the same pathological stage, with different characteristics of clinical onset and different individual response to treatment. Moreover, patients with stage II and III CRC are shown to have different prognoses, particularly those who receive adjuvant chemotherapy, with 5-year overall survival (OS) ranging between 50% and 90% [9].

Adjuvant chemotherapy is strongly indicated in stage III disease, which is associated with a reduction of the relative risk of death of 33%, and an absolute survival benefit of 5–10% [10]. In stage III, the use of oxaliplatin in addition to fluoropyrimidines yields a further significant advantage of about 5% in terms of disease-free survival (DFS) and OS. Conversely, the therapeutic indication in patients with stage II CRC is controversial, as treatment with 5-Fluorouracil has an absolute benefit of 3–4% [11,12]. In patients with clinicopathologically high-risk stage II disease [13], decision-making around adjuvant chemotherapy treatment needs to be carefully evaluated and discussed, considering also recurrence risk factors such as baseline carcinoembryonic antigen and vascular invasion [7]. There is no evidence to support the use of adjuvant chemotherapy in stage I disease. Considering all the above mentioned data, identifying patients who are most likely to benefit from adjuvant chemotherapy and preventing the other patients from futile treatments and unnecessary exposure to toxicity is crucial in stage II disease.

Early detection of disease relapse is extremely relevant in CRC, as radical surgical intervention in patients with oligometastatic CRC can achieve a proven survival benefit. Therefore, early detection of relapse could potentially increase cure rates. Postoperative surveillance with clinical, radiological, and markers examination is often unable to identify early metastatic disease and/or postoperative minimal residual disease. Based on these considerations, improved risk stratification tools are required to reduce the number of patients treated unnecessarily.

Metabolomics is defined as the comprehensive measurement of the ensemble of metabolites present in a biological specimen, the so-called metabolome [14]. Metabolites represent, at the same time, the downstream output of the genome, transcriptome, and proteome, as well as the upstream input from various exogenous factors such as environment, lifestyle, diet, and drug administration [15]. In contrast to genomics, which indicates what might happen, metabolomic profiling/phenotyping captures what is actually happening in the body, and for this reason, in the last few years, metabolomics has been extensively applied in biomedical research [16–22].

Several relevant efforts to improve risk stratification in CRC have been made in the past years, considering mismatch repair (MMR) status, as well as BRAF and KRAS mutations, and the presence of tumor-derived circulating DNA [23,24]. Metabolomics has also emerged as a technique capable of contributing significantly in this setting [25–31]. Some of us have shown, in a cohort of elderly patients, that nuclear magnetic resonance (NMR)-based metabolomics can discriminate between early and metastatic CRC. This approach may be a useful tool to build a prognostic model capable of assessing the likelihood of cancer relapse, based on the degree to which a serum fingerprint derived from a patient with early disease resembles that of a metastatic patient [13].

The study presented here explores the variations among preoperative and postoperative metabolomic serum fingerprints of CRC patients, obtained via NMR spectroscopy (Figure 1); moreover, for the first time, to the best of our knowledge, possible associations between pre/post-surgery metabolic variations and cancer relapse are examined.

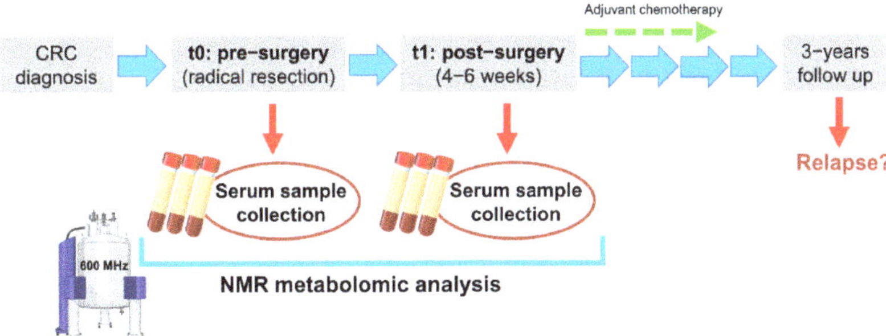

Figure 1. Study design.

2. Materials and Methods

2.1. Study Cohort

From June 2017 to August 2018, we prospectively enrolled 41 patients with histologically diagnosed CRC, who were treated as per standard clinical practice at the Prato Hospital. All patients enrolled met the following inclusion criteria: (i) female or male patients with radically operable heteroplasia of the colon/rectum (stage I, II, III); (ii) Eastern Cooperative Oncology Group Scale of Performance Status (ECOG PS) 0–1; (iii) patients of age $\geq$ 18 years. For all patients enrolled, the following data were collected: (i) demographic data; (ii) clinical and histological characterization of the tumor; (iii) any other clinical information useful for the study (i.e., comorbidities, drug treatments).

All patients signed informed consent before entry into the study. The present study complies with the 1964 Declaration of Helsinki and its later amendments and received the approval by the local ethics committee (Comitato Etico Regione Toscana—Area Vasta Centro, study number: 10208_bio).

2.2. Samples Collection

Serum samples were collected and stored following standard operating procedures validated at international level [32]. Two $\times$ 10 mL of overnight fasting peripheral blood were collected for each patient at the two timepoints (t0: before the radical tumor resection; t1: 4–6 weeks after surgery before the start of any adjuvant treatment) in serum vacutainer and processed within one hour from phlebotomy. After clot formation at room temperature, tubes were centrifuged at 1600 RCF for 10 min at 4 °C. Then, serum aliquots of 1 mL (labelled with an anonymized code) were immediately frozen at −80 °C, pending NMR analysis.

2.3. NMR Analysis

2.3.1. Acquisition of NMR Data

All NMR spectra were acquired using a Bruker 600 MHz spectrometer (Bruker BioSpin, Rheinstetten, Germany) operating at 600.13 MHz proton Larmor frequency, equipped with an automatic refrigerated (6 °C) sample changer (SampleJet, Bruker BioSpin). Temperature stabilization (approximately 0.1 K at the sample) was obtained using a BTO 2000 thermocouple. Before NMR acquisition, to equilibrate temperature at 310 K, each sample was maintained inside the NMR probe head for at least 300 s. The spectrometer was calibrated daily, before any measurement, following strict standard operation procedures [33] to ensure high spectral quality and reproducibility.

Serum samples contain low molecular weight metabolites as well as high molecular weight macromolecules; for this reason, three different pulse sequences were used to enable the selective detection of the different serum molecular components: a 1D spin echo Carr–Purcell–Meiboom–Gill sequence (CPMG) was used to selectively detect signals of low molecular weight metabolites, and a 1D diffusion-edited pulse sequence was used to selectively acquire the signals of high molecular weight components (i.e., lipids, lipoproteins, proteins). Moreover, a 1D nuclear Overhauser effect spectroscopy pulse sequence (NOESY) was applied to detect signals of all molecules present in concentrations above the NMR detection limit.

A detailed description of sample preparation procedures, instrument configuration, and NMR parameters setting can be retrieved from our previous publication [16].

2.3.2. Spectral Processing

Before applying Fourier transform, free induction decays were multiplied by an exponential function equivalent to a 0.3 Hz line-broadening factor. Using automated routine of TopSpin 3.6 (Bruker BioSpin), Fourier-transformed spectra were corrected for phase and baseline distortions, and NOESY and CPMG spectra were also calibrated at the anomeric glucose ^{1}H doublet at δ 5.24 ppm. Each 1D spectrum in the range between 0.2 and 10.0 ppm was segmented into chemical shift bins of 0.02 ppm, and the corresponding spectral areas were integrated using AssureNMR software (Bruker BioSpin). The spectral region containing residual water signal (δ 5.12–4.38 ppm) was removed, and the dimension of the system was reduced to 453 bins.

2.4. Statistical Analysis

All data analysis was executed in the "R" statistical environment [34]. Multivariate analysis was performed on binned spectra without any a priori knowledge of the metabolites present. Multilevel partial least square analysis (mPLS) [35,36] was performed to obtain data reduction (R script developed in-house). Support vector machine [37] applied on the first nine mPLS components was used for classification purposes. Models were evaluated by means of 100 cycles of a Monte Carlo cross-validation scheme (in-house-developed R script). In brief, 90% of the pairs of data, selected at random at each iteration, were used as a training set to build the model. Then, the remaining 10% was tested, and sensitivity, specificity, and accuracy (calculated according to the standard definitions) were assessed.

Univariate analysis was conducted directly on the spectral regions associated with the metabolites/lipoproteins present in all serum samples at concentrations above the detection limit (>1 μM). Metabolites and lipoprotein-related parameters were identified and quantified using the Bruker IVDr quantification platform [38]. Metabolites whose levels were lower than the limit of quantification (LOQ) were imputed with half the LOQ (Table S1). Nonparametric Wilcoxon signed-rank test was used to infer intraindividual differences between the two timepoints. The p-values were adjusted for multiple testing using the false discovery rate (FDR) procedure with Benjamini–Hochberg [39] correction at $\alpha = 0.05$. Wilcoxon rank-sum test was used to infer differences between metabolites/lipoproteins of free-from-disease and relapsed CRC patients. The p-values were not adjusted for multiple testing because the group of relapse patients is small, and therefore the correction would be too severe, increasing the risk of missing promising biomarkers. However, we are aware that this could increase the risk of a type I error.

Univariate analysis on clinical data was performed using the Fisher test for categorical variables and the ANOVA test for continuous variables. Polyserial correlations between ordinal clinical variables (pT, N, grade, stage, ECOG PS) and metabolites were calculated using the function "polyserial" (R package "polycor"). Point-biserial correlations between dichotomous clinical variables (tumor localization, sex) and metabolites were calculated using the function "biserial.cor" (R package "ltm").

3. Results

3.1. Characteristics of Enrolled Patients

Forty-one patients were enrolled in the study (21 female and 20 male). The median age was 73 years (Table 1).

Table 1. Descriptive statistics of enrolled CC patients at the time of analysis.

		Whole Sample (N = 41)	Stratified by Progression Status			Stratified by Chemotherapy Treatment			
			Not Relapsed (N = 33)	Relapsed (N = 8)	p-Value	Capecitabine (N = 9)	XELOX (N = 10)	No CT (N = 22)	p-Value
Age at study entry	Median (min; max)	73 (51;92)	71 (51;92)	78 (68;86)	0.032	77 (68;86)	65 (51;72)	78 (54;92)	0.001
Sex	F	21 (51%)	18 (55%)	3 (38%)	0.454	2 (22%)	7 (70%)	12 (55%)	0.109
	M	20 (49%)	15 (45%)	5 (62%)		7 (78%)	3 (30%)	10 (45%)	
ECOG PS	PS 0	29 (71%)	23 (70%)	6 (75%)	1	7 (78%)	10 (100%)	12 (55%)	0.102
	PS 1	8 (20%)	7 (21%)	1 (12%)		1 (11%)	0 (0%)	7 (32%)	
	PS 2	4 (10%)	3 (9%)	1 (12%)		1 (11%)	0 (0%)	3 (14%)	
pT	pT1	6 (15%)	6 (18%)	0 (0%)	0.086	0 (0%)	1 (10%)	5 (23%)	0.477
	pT2	8 (20%)	8 (24%)	0 (0%)		1 (11%)	1 (10%)	6 (27%)	
	pT3	23 (56%)	17 (52%)	6 (75%)		7 (78%)	7 (70%)	9 (41%)	
	pT4	4 (10%)	2 (6%)	2 (25%)		1 (11%)	1 (10%)	2 (9%)	
N	N0	24 (59%)	23 (70%)	1 (12%)	0.005	3 (33%)	3 (30%)	18 (82%)	0.005
	N+	17 (41%)	10 (30%)	7 (88%)		6 (67%)	7 (70%)	4 (18%)	
Stage risk	Stage I	11 (27%)	11 (33%)	0 (0%)	0.035	0 (0%)	0 (0%)	11 (50%)	0.002
	Stage II Low risk	2 (5%)	2 (6%)	0 (0%)		0 (0%)	0 (0%)	2 (9%)	
	Stage II High risk	11 (27%)	10 (30%)	1 (12%)		3 (33%)	3 (30%)	5 (23%)	
	Stage III	17 (41%)	10 (30%)	7 (88%)		6 (67%)	7 (70%)	4 (18%)	
Grading	G1	2 (5%)	1 (3%)	1 (12%)	0.168	1 (11%)	1 (10%)	0 (0%)	0.205
	G2	19 (48%)	17 (53%)	2 (25%)		2 (22%)	4 (40%)	13 (62%)	
	G3	17 (42%)	13 (41%)	4 (50%)		5 (56%)	5 (50%)	7 (33%)	
	G4	2 (5%)	1 (3%)	1 (12%)		1 (11%)	0 (0%)	1 (5%)	
	NA	1	1	0		0	0	1	
Localization	Left-sided	13 (32%)	12 (36%)	1 (12%)	0.398	3 (33%)	6 (60%)	4 (18%)	0.07
	Right-sided	28 (68%)	21 (64%)	7 (88%)		6 (67%)	4 (40%)	18 (82%)	
Comorbidities	No com.	13 (32%)	8 (24%)	5 (62%)	0.111	4 (44%)	3 (30%)	6 (27%)	0.519
	No vascular com.	8 (20%)	8 (24%)	0 (0%)		0 (0%)	3 (30%)	5 (23%)	
	Vascular com.	20 (49%)	17 (52%)	3 (38%)		5 (56%)	4 (40%)	11 (50%)	
MSI	Instable	1 (11%)	1 (14%)	0 (0%)	1	0 (0%)	0 (0%)	1 (33%)	0.278
	MSI	1 (11%)	1 (14%)	0 (0%)		0 (0%)	0 (0%)	1 (33%)	
	Stable	7 (78%)	5 (71%)	2 (100%)		2 (100%)	4 (100%)	1 (33%)	
	NA	32	26	6		7	6	19	
CDX2	Positive	1 (100%)	0 (0%)	1 (100%)	-	1 (100%)	0 (0%)	0 (0%)	-
	NA	40	33	7		8	10	22	
KRAS	Mutated	5 (29%)	1 (11%)	4 (50%)	0.131	3 (50%)	0 (0%)	2 (67%)	0.042
	WT	12 (71%)	8 (89%)	4 (50%)		3 (50%)	8 (100%)	1 (33%)	
	NA	24	24	0		3	2	19	
NRAS	WT	13 (100%)	8 (100%)	5 (100%)	-	4 (100%)	8 (100%)	1 (100%)	-
	NA	28	25	3		5	2	21	
BRAF	Mutated	4 (24%)	3 (33%)	1 (12%)	0.576	1 (17%)	3 (38%)	0 (0%)	0.461
	WT	13 (76%)	6 (67%)	7 (88%)		5 (83%)	5 (62%)	3 (100%)	
	NA	24	24	0		3	2	19	

ECOG PS: Eastern Cooperative Oncology Group Performance Status; pT: primary tumor size; N: regional lymph nodes; MSI: microsatellite instability.

Most of the enrolled patients had a good Eastern Cooperative Oncology Group (ECOG) performance status (PS), with 29 patients (71%) having a PS 0. However, over one half of the patients (n = 38; 69%) had comorbidity, of which 20 patients had vascular comorbidity.

By inclusion criteria, all patients have early-stage disease: 11 patients (27%) with stage I, 13 patients (32%) stage II, and 17 patients (41%) stage III. In particular, six patients had a pT1 (5 N0 e 1 N+), eight patients had a pT2 (6 N0 and 2 N+), 23 patients had a pT3 (18 N+), and four patients had a pT4 (3 N0 and 1 N+).

Regarding the 13 patients with stage II, two were at low risk and 11 at high risk for the presence of lymphovascular invasion, T4 or G3–4.

The majority of tumors had intermediate (G2; 48%: N = 19) or high (G3–G4; 47% N = 19) histologic grading, while G1 accounted for 5% of tumors in this population (N = 2). A total of 13 patients had left CRC and 28 right CRC (Table 1).

Half of the patients (46%; n = 19) received adjuvant chemotherapy, in accordance with clinical stage of disease. Nine patients received fluoropyrimidine monotherapy and 10 patients received polychemotherapy with oxaliplatin and fluoropyrimidine. Six out of eleven patients at stage II at high risk received adjuvant treatment; the rest of them did not receive chemotherapy for age or comorbidity

Thirteen out of the 17 patients with stage III disease received adjuvant treatment according to tumor stage. At the last follow-up, 19% (n = 8) of patients had disease relapse (Table 1). As expected, the patients with relapse had a history of stage III disease or stage II at high risk.

3.2. *Effects of Surgery on the Metabolome of CRC Patients*

The mPLS analysis was performed to assess intraindividual variations between t0 and t1 in the metabolomic fingerprints of CRC patients. The results obtained show significant differential clustering, with optimal separation of the two timepoints using each type of NMR spectra acquired, namely CPMG, NOESY, and DIFFUSION (Figure 2). All models classify t0 and t1 samples with an accuracy in the range 82–85%, and the best results were obtained using NOESY spectra. These data indicate that both low molecular weight metabolites and high molecular weight macromolecules (i.e., lipoproteins, proteins) contribute to the discrimination.

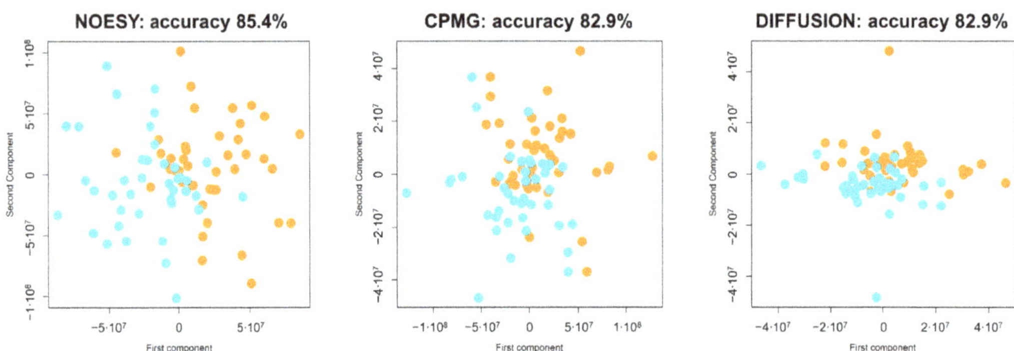

Figure 2. Score plots of the first two components of the mPLS models calculated using each of the three typologies of NMR spectra acquired: CPMG; NOESY; diffusion-edited. Discrimination accuracy of each model is reported. Each dot represents an NMR spectrum; dots are colored as follows: t0—orange, t1—turquoise. The first component mainly describes the differences between t0 and t1. The second component mainly reports the within-subject variation.

From univariate analysis emerges that after surgery there is a significant increase of pyruvate, HDL cholesterol, HDL phospholipids, HDL Apo-A1, and HDL Apo-A2 (Figure 3). Moreover, after surgery we observed a significant decrement of acetone, 3-hydroxybutyrate, LDL-Chol/HDL-Chol ratio, and Apo-A1/Apo-B100 ratio (Figure 3). Furthermore, several lipoprotein-related subfractions were shown to be significantly altered between t0 and t1 (Figure S1). These data point to a relevant rearrangement of the metabolic pathways related to lipoproteins, ketone bodies, and energy metabolism.

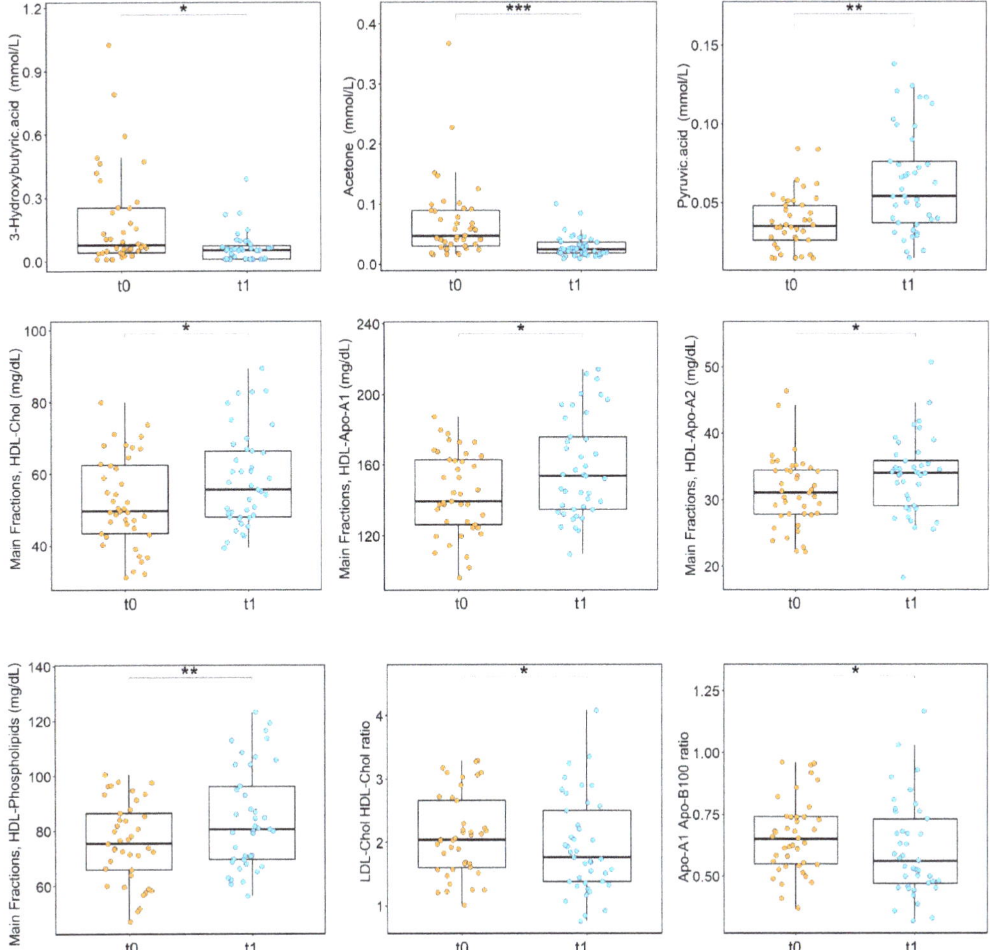

Figure 3. Boxplots of the statistically significant metabolites and lipoproteins-related parameters discriminating CRC patients at t0 (orange) and t1 (turquoise); *p*-values obtained using Wilcoxon signed-rank test and adjusted for FDR are reported. *** $p < 0.001$; ** $p < 0.01$; * $p < 0.05$.

3.3. Associations between Metabolome Variations after Surgery and Cancer Relapse

Eight out of the 41 CRC patients enrolled in the present study developed cancer relapse in the three years after diagnosis. We hypothesized that different changes in preoperative and postoperative metabolomic serum profiles could be predictive of patients' prognosis. To explore this hypothesis, the difference between each metabolite/lipoprotein-related parameter at t1 and t0 was calculated, and each resulting difference analyzed via univariate approaches to underline possible divergent behavior in free-from-disease and relapsed CRC patients. Postoperative, relapsed CRC patients show a significant increase of pyruvate, HDL Apo-A1, HDL Apo-A2, HDL cholesterol, HDL free cholesterol, and HDL phospholipids, and a significant decrease of Apo-A1 Apo-B100 ratio, VLDL-5 cholesterol, VLDL-5 free cholesterol, and VLDL-5 phospholipids (Figure 4).

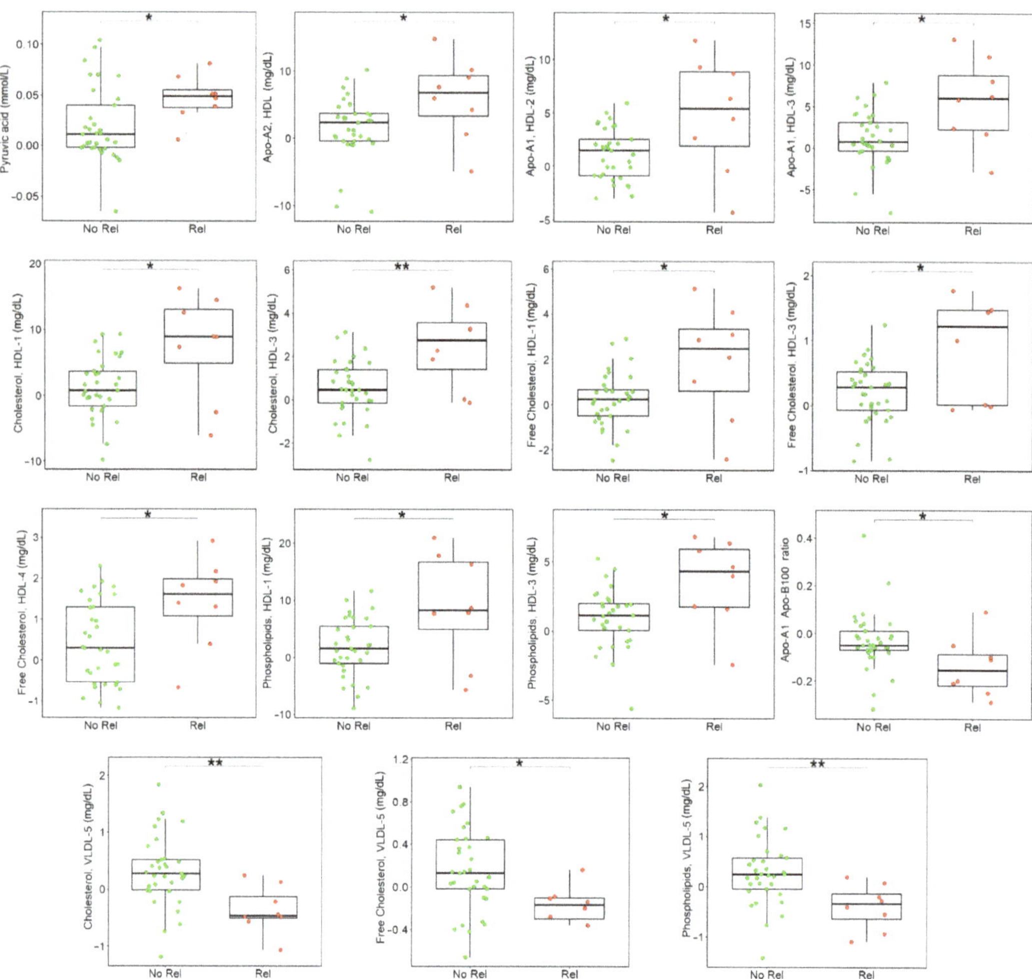

Figure 4. Boxplots of the differences between t1 and t0 discriminating free-from-disease (green) and relapsed (red) patients, only statistically significant metabolites and lipoproteins-related parameters are reported; *p*-values obtained using Wilcoxon signed-rank test are reported. ** $p < 0.01$; * $p < 0.05$.

3.4. Associations between Metabolites and Clinical Variables

Possible associations between metabolites/lipoproteins (main fractions) and clinical variables were investigated. Results are reported in Table S2.

Glycine and histidine showed statistically significant correlations with tumor size. Tyrosine correlates with tumor stage and regional lymph nodal spread (N). N also correlates with isoleucine, Apo-A1, and Apo-A2. Tumor localization (left or right colon) shows correlations with acetone, cholesterol, LDL cholesterol, and Apo-B100. Interestingly a panel of eight metabolic variables (N,N-Dimethylglycine, valine, dimethylsulfone, triglycerides, cholesterol, LDL cholesterol, Apo-A2, Apo-B100) correlates with the Eastern Cooperative Oncology Group Scale of Performance Status. Moreover, as expected, sex shows significant correlations with several metabolites/lipoproteins: creatine, creatinine, glutamine, glycine, isoleucine, leucine, formic acid, cholesterol, LDL cholesterol, HDL cholesterol, Apo-A1, Apo-A2, and Apo-B100. Of note, none of the examined metabolic features show significant correlation with tumor grade.

4. Discussion

The primary option for the treatment of colorectal cancer is surgery. Adjuvant chemotherapy is strongly indicated in stage III disease and in stage II patients at high risk of relapse. Whereas, in low-risk stage II disease decision-making around adjuvant chemotherapy must be carefully evaluated. At present, postoperative surveillance via clinical, radiological and biomarkers examination often cannot identify early metastatic disease and/or postoperative micrometastatic residual disease.

Based on these considerations, especially in stage II disease, improved risk-stratification tools are required to identify those patients who are most likely to benefit from adjuvant chemotherapy and need to be followed up more closely after surgery to timely detect systemic recurrence. On the other hand, accurate stratification instruments could prevent low-risk patients from unnecessary treatment and possible mild-to-severe adverse reactions.

The analysis described in the present research article shows for the first time, to the best of our knowledge, the metabolomic variations among preoperative and postoperative NMR-based serum fingerprint of CRC patients. Furthermore, metabolomics as novel approach for risk-stratification in CRC setting was evaluated by studying differences between pre- and postoperative serum samples of each patient enrolled. With this innovative approach, each patient in the study population acts as his/her own control, thus eliminating noise from interindividual variability.

Our data demonstrate that metabolomics profiles are influenced by the presence or absence of the cancerous mass. Indeed, the mPLS models calculated using each of the three NMR spectra acquired (namely, CPMG, NOESY, and DIFFUSION) show high discrimination accuracies (range 82–85%). This evidence poses an important question in terms of future study design, since sample collection when the tumor was still in place or after resection can significantly impact on metabolomic data.

From the univariate analysis, it emerges that after surgery, there is a significant increase of pyruvate, HDL cholesterol, HDL phospholipids, HDL Apo-A1, and HDL Apo-A2. Moreover, we observed, postoperative, a significant decrement of acetone, 3-hydroxybutyrate, LDL-Chol/HDL-Chol ratio, and Apo-A1/Apo-B100 ratio. These data point to a relevant rewiring of the metabolic pathways associated to lipoproteins, ketone bodies, and energy metabolism.

Depletion of pyruvate and increase of ketone bodies has been observed in sera of metastatic CRC patients with respect to healthy controls, and this evidence has been associated with an altered energy metabolism, probably reflecting an increased gluconeogenesis and fatty acid oxidation [31]. It is interesting to note that in our dataset, these three metabolites show trend inversions after surgery.

Our data show an increase of several HDL-Chol and a decrease of LDL-Chol lipoprotein-related parameters post-surgery. This may be explained by the fact that, after cancer resection, an improvement in the inflammatory status of the gut is achieved, allowing for an improved lipid metabolism and lipid assimilation in the absence of the tumor.

Strikingly, despite the low number of recurrence events registered, it is peculiar that the difference in HDL-Chol is particularly marked in relapsed patients and is coupled with a decrease of VLDL-Chol. It has been observed that in colorectal cancerous tissue, the levels of cholesterol and triglycerides were reduced and HDL-Cholesterol level increased, indicating that CRC development destroys the physiological balance of lipids and lipoproteins, leading to lipid metabolic disorders [40]. Preclinical and clinical studies have already investigated the role of cholesterol in CRC progression; however, a clear understanding of the molecular mechanism linking these two entities is still lacking [40,41].

In conclusion, our results show that surgery can affect the metabolomic and lipidomic profiles of CRC patients and they point to possible associations between these metabolic changes and cancer recurrence. This study is based on a small population of CRC patients in which a very limited number of recurrence events are present; therefore, at present, results are only speculative and require further confirmation. In order to validate these findings in a general population, we are conducting a multicentric prospective trial focused on high-risk stage disease, the LIquid BIopsy and METabolomics in colon cancer (LIBIMET) study. LIBIMET aims primarily at redefining the risk of relapse in patients with high-risk, early-stage colon cancer by combining of ctDNA and serum metabolomics.

5. Conclusions

Taken together, the data here presented highlight the notion that CRC can induce metabolic changes that are reflected at a systemic level and can be detected in serum. This evidence suggests that our approach aimed at detecting micrometastatic CRC by assessing its metabolomic fingerprint in serum is correct, and that this may be exploited for biomarker-oriented research to contribute towards better management of colorectal cancer.

Supplementary Materials: The following are available online at https://www.mdpi.com/article/10.3390/app112311120/s1. Table S1: Data completeness for the different metabolites quantified in the serum samples analyzed via NMR. LOQ = limit of quantification. Table S2: Correlation between clinical data and metabolites. Correlation coefficients and p-values are reported in table. Figure S1: Boxplots of the statistically significant lipoproteins-related parameters discriminating of CRC patients at t0 (orange) and t1 (turquoise); p-values obtained using Wilcoxon signed-rank test and adjusted for FDR are reported.

Author Contributions: Study conception and design, E.M., S.D.D., C.L., L.T. and L.B.; patient enrolment and management: E.M., S.D.D., V.C., S.C., M.B. (Maddalena Baraghini) and A.G.; collection of clinical data and serum samples: E.M., S.D.D., C.B., M.B. (Maddalena Baraghini), V.C., A.P., S.C., M.B. (Matteo Benelli), A.G. and F.D.M.; NMR analysis: A.V.; statistical analysis, biostatistics, and computational analysis, A.V., C.B., M.B. (Matteo Benelli), D.R. and L.T.; results interpretation, A.V., E.M., S.D.D., L.M., I.M., C.L., L.T. and L.B.; writing—original draft preparation: A.V. and E.M.; writing—review and editing: A.V., E.M., S.D.D., L.M., C.B., M.B. (Matteo Benelli), V.C., A.P., S.C., M.B. (Maddalena Baraghini), A.G., F.D.M., D.R., I.M., C.L., L.T. and L.B.; supervision, C.L, L.T. and L.B. All authors have read and agreed to the published version of the manuscript.

Funding: This research received no external funding.

Institutional Review Board Statement: The study was conducted according to the guidelines of the Declaration of Helsinki and approved by the local ethics committee (Comitato Etico Regione Toscana—Area Vasta Centro, study number: 10208_bio).

Informed Consent Statement: Written informed consent was obtained from all subjects involved in the study.

Data Availability Statement: Data and R script are available from the corresponding authors upon reasonable request.

Acknowledgments: In memoriam of Angelo Di Leo who passed away on 13 June 2021, while this work was being completed. The authors acknowledge the Fondazione Pitigliani per la lotta contro i tumori ONLUS for its support. The authors acknowledge Instruct-ERIC, a Landmark ESFRI project, and specifically the CERM/CIRMMP Italy Centre. Alessia Vignoli was supported by an AIRC fellowship for Italy.

Conflicts of Interest: The authors declare no conflict of interest.

References

1. NCCN Guidelines for Colon Cancer 2021. Available online: https://www.nccn.org/guidelines/guidelines-detail (accessed on 5 November 2021).
2. Siegel, R.L.; Miller, K.D.; Jemal, A. Cancer Statistics, 2020. *CA Cancer J. Clin.* **2020**, *70*, 7–30. [CrossRef] [PubMed]
3. Sung, H.; Ferlay, J.; Siegel, R.L.; Laversanne, M.; Soerjomataram, I.; Jemal, A.; Bray, F. Global Cancer Statistics 2020: GLOBOCAN Estimates of Incidence and Mortality Worldwide for 36 Cancers in 185 Countries. *CA Cancer J. Clin.* **2021**, *71*, 209–249. [CrossRef] [PubMed]
4. AIOM: Linee Guida Tumori del Colon 2020. Available online: https://www.aiom.it/linee-guida-aiom-2020-tumori-del-colon/ (accessed on 5 November 2021).
5. Reinert, T.; Schøler, L.V.; Thomsen, R.; Tobiasen, H.; Vang, S.; Nordentoft, I.; Lamy, P.; Kannerup, A.-S.; Mortensen, F.V.; Stribolt, K.; et al. Analysis of Circulating Tumour DNA to Monitor Disease Burden Following Colorectal Cancer Surgery. *Gut* **2016**, *65*, 625–634. [CrossRef] [PubMed]

6. Guraya, S.Y. Pattern, Stage, and Time of Recurrent Colorectal Cancer after Curative Surgery. *Clin. Colorectal Cancer* **2019**, *18*, e223–e228. [CrossRef]
7. Hall, M.J.; Morris, A.M.; Sun, W. Precision Medicine Versus Population Medicine in Colon Cancer: From Prospects of Prevention, Adjuvant Chemotherapy, and Surveillance. *Am. Soc. Clin. Oncol. Educ. Book* **2018**, *38*, 220–230. [CrossRef] [PubMed]
8. Dienstmann, R.; Mason, M.J.; Sinicrope, F.A.; Phipps, A.I.; Tejpar, S.; Nesbakken, A.; Danielsen, S.A.; Sveen, A.; Buchanan, D.D.; Clendenning, M.; et al. Prediction of Overall Survival in Stage II and III Colon Cancer beyond TNM System: A Retrospective, Pooled Biomarker Study. *Ann. Oncol.* **2017**, *28*, 1023–1031. [CrossRef]
9. SEER Cancer Statistics Review, 1975–2017. Available online: https://seer.cancer.gov/csr/1975_2017/index.html (accessed on 30 March 2021).
10. Renfro, L.A.; Grothey, A.; Xue, Y.; Saltz, L.B.; André, T.; Twelves, C.; Labianca, R.; Allegra, C.J.; Alberts, S.R.; Loprinzi, C.L.; et al. ACCENT-Based Web Calculators to Predict Recurrence and Overall Survival in Stage III Colon Cancer. *J. Natl. Cancer Inst.* **2014**, *106*, dju333. [CrossRef]
11. Benson, A.B.; Schrag, D.; Somerfield, M.R.; Cohen, A.M.; Figueredo, A.T.; Flynn, P.J.; Krzyzanowska, M.K.; Maroun, J.; McAllister, P.; Van Cutsem, E.; et al. American Society of Clinical Oncology Recommendations on Adjuvant Chemotherapy for Stage II Colon Cancer. *J. Clin. Oncol.* **2004**, *22*, 3408–3419. [CrossRef]
12. Kumar, A.; Kennecke, H.F.; Renouf, D.J.; Lim, H.J.; Gill, S.; Woods, R.; Speers, C.; Cheung, W.Y. Adjuvant Chemotherapy Use and Outcomes of Patients with High-Risk versus Low-Risk Stage II Colon Cancer. *Cancer* **2015**, *121*, 527–534. [CrossRef]
13. Di Donato, S.; Vignoli, A.; Biagioni, C.; Malorni, L.; Mori, E.; Tenori, L.; Calamai, V.; Parnofiello, A.; Di Pierro, G.; Migliaccio, I.; et al. A Serum Metabolomics Classifier Derived from Elderly Patients with Metastatic Colorectal Cancer Predicts Relapse in the Adjuvant Setting. *Cancers* **2021**, *13*, 2762. [CrossRef] [PubMed]
14. Nicholson, J.K.; Lindon, J.C. Systems Biology: Metabonomics. *Nature* **2008**, *455*, 1054–1056. [CrossRef]
15. Vignoli, A.; Risi, E.; McCartney, A.; Migliaccio, I.; Moretti, E.; Malorni, L.; Luchinat, C.; Biganzoli, L.; Tenori, L. Precision Oncology via NMR-Based Metabolomics: A Review on Breast Cancer. *Int. J. Mol. Sci.* **2021**, *22*, 4687. [CrossRef]
16. Vignoli, A.; Ghini, V.; Meoni, G.; Licari, C.; Takis, P.G.; Tenori, L.; Turano, P.; Luchinat, C. High-Throughput Metabolomics by 1D NMR. *Angew. Chem. Int. Ed.* **2019**, *58*, 968–994. [CrossRef] [PubMed]
17. Wishart, D.S. Emerging Applications of Metabolomics in Drug Discovery and Precision Medicine. *Nat. Rev. Drug Discov.* **2016**, *15*, 473–484. [CrossRef] [PubMed]
18. Vignoli, A.; Tenori, L.; Giusti, B.; Valente, S.; Carrabba, N.; Baizi, D.; Barchielli, A.; Marchionni, N.; Gensini, G.F.; Marcucci, R.; et al. Differential Network Analysis Reveals Metabolic Determinants Associated with Mortality in Acute Myocardial Infarction Patients and Suggests Potential Mechanisms Underlying Different Clinical Scores Used To Predict Death. *J. Proteome Res.* **2020**, *19*, 949–961. [CrossRef] [PubMed]
19. Zhang, L.; Zhu, B.; Zeng, Y.; Shen, H.; Zhang, J.; Wang, X. Clinical Lipidomics in Understanding of Lung Cancer: Opportunity and Challenge. *Cancer Lett.* **2020**, *470*, 75–83. [CrossRef] [PubMed]
20. Pietzner, M.; Stewart, I.D.; Raffler, J.; Khaw, K.-T.; Michelotti, G.A.; Kastenmüller, G.; Wareham, N.J.; Langenberg, C. Plasma Metabolites to Profile Pathways in Noncommunicable Disease Multimorbidity. *Nat. Med.* **2021**, *27*, 471–479. [CrossRef] [PubMed]
21. Vignoli, A.; Paciotti, S.; Tenori, L.; Eusebi, P.; Biscetti, L.; Chiasserini, D.; Scheltens, P.; Turano, P.; Teunissen, C.; Luchinat, C.; et al. Fingerprinting Alzheimer's Disease by 1H Nuclear Magnetic Resonance Spectroscopy of Cerebrospinal Fluid. *J. Proteome Res.* **2020**, *19*, 1696–1705. [CrossRef] [PubMed]
22. Vignoli, A.; Tenori, L.; Giusti, B.; Takis, P.G.; Valente, S.; Carrabba, N.; Balzi, D.; Barchielli, A.; Marchionni, N.; Gensini, G.F.; et al. NMR-Based Metabolomics Identifies Patients at High Risk of Death within Two Years after Acute Myocardial Infarction in the AMI-Florence II Cohort. *BMC Med.* **2019**, *17*, 3. [CrossRef]
23. Auclin, E.; Zaanan, A.; Vernerey, D.; Douard, R.; Gallois, C.; Laurent-Puig, P.; Bonnetain, F.; Taieb, J. Subgroups and Prognostication in Stage III Colon Cancer: Future Perspectives for Adjuvant Therapy. *Ann. Oncol.* **2017**, *28*, 958–968. [CrossRef]
24. Copija, A.; Waniczek, D.; Witkoś, A.; Walkiewicz, K.; Nowakowska-Zajdel, E. Clinical Significance and Prognostic Relevance of Microsatellite Instability in Sporadic Colorectal Cancer Patients. *Int. J. Mol. Sci.* **2017**, *18*, 107. [CrossRef] [PubMed]
25. Nannini, G.; Meoni, G.; Amedei, A.; Tenori, L. Metabolomics Profile in Gastrointestinal Cancers: Update and Future Perspectives. *World J. Gastroenterol.* **2020**, *26*, 2514–2532. [CrossRef] [PubMed]
26. Ma, Y.; Zhang, P.; Wang, F.; Liu, W.; Yang, J.; Qin, H. An Integrated Proteomics and Metabolomics Approach for Defining Oncofetal Biomarkers in the Colorectal Cancer. *Ann. Surg.* **2012**, *255*, 720–730. [CrossRef]
27. Nishiumi, S.; Kobayashi, T.; Ikeda, A.; Yoshie, T.; Kibi, M.; Izumi, Y.; Okuno, T.; Hayashi, N.; Kawano, S.; Takenawa, T.; et al. A Novel Serum Metabolomics-Based Diagnostic Approach for Colorectal Cancer. *PLoS ONE* **2012**, *7*, e40459. [CrossRef]
28. Qiu, Y.; Cai, G.; Zhou, B.; Li, D.; Zhao, A.; Xie, G.; Li, H.; Cai, S.; Xie, D.; Huang, C.; et al. A Distinct Metabolic Signature of Human Colorectal Cancer with Prognostic Potential. *Clin. Cancer Res.* **2014**, *20*, 2136–2146. [CrossRef]
29. Farshidfar, F.; Weljie, A.M.; Kopciuk, K.; Buie, W.D.; Maclean, A.; Dixon, E.; Sutherland, F.R.; Molckovsky, A.; Vogel, H.J.; Bathe, O.F. Serum Metabolomic Profile as a Means to Distinguish Stage of Colorectal Cancer. *Genome Med.* **2012**, *4*, 42. [CrossRef] [PubMed]
30. Farshidfar, F.; Weljie, A.M.; Kopciuk, K.A.; Hilsden, R.; McGregor, S.E.; Buie, W.D.; MacLean, A.; Vogel, H.J.; Bathe, O.F. A Validated Metabolomic Signature for Colorectal Cancer: Exploration of the Clinical Value of Metabolomics. *Br. J. Cancer* **2016**, *115*, 848–857. [CrossRef]

31. Bertini, I.; Cacciatore, S.; Jensen, B.V.; Schou, J.V.; Johansen, J.S.; Kruhøffer, M.; Luchinat, C.; Nielsen, D.L.; Turano, P. Metabolomic NMR Fingerprinting to Identify and Predict Survival of Patients with Metastatic Colorectal Cancer. *Cancer Res.* **2012**, *72*, 356–364. [CrossRef] [PubMed]
32. ISO/DIS 23118 Molecular In Vitro Diagnostic Examinations—Specifications for Pre-Examination Processes in Metabolomics in Urine, Venous Blood Serum and Plasma. Available online: https://www.iso.org/obp/ui/#iso:std:iso:23118:ed-1:v1:en (accessed on 4 June 2021).
33. Bruzzone, C.; Bizkarguenaga, M.; Gil-Redondo, R.; Diercks, T.; Arana, E.; García de Vicuña, A.; Seco, M.; Bosch, A.; Palazón, A.; San Juan, I.; et al. SARS-CoV-2 Infection Dysregulates the Metabolomic and Lipidomic Profiles of Serum. *iScience* **2020**, *23*, 101645. [CrossRef]
34. R Core Team. *R: A Language and Environment for Statistical Computing*; R Foundation for Statistical Computing: Vienna, Austria, 2014.
35. van Velzen, E.J.J.; Westerhuis, J.A.; van Duynhoven, J.P.M.; van Dorsten, F.A.; Hoefsloot, H.C.J.; Jacobs, D.M.; Smit, S.; Draijer, R.; Kroner, C.I.; Smilde, A.K. Multilevel Data Analysis of a Crossover Designed Human Nutritional Intervention Study. *J. Proteome Res.* **2008**, *7*, 4483–4491. [CrossRef]
36. Westerhuis, J.A.; van Velzen, E.J.; Hoefsloot, H.C.; Smilde, A.K. Multivariate Paired Data Analysis: Multilevel PLSDA versus OPLSDA. *Metabolomics* **2010**, *6*, 119–128. [CrossRef]
37. Cortes, C.; Vapnik, V. Support-Vector Networks. *J. Mach. Learn. Res.* **1995**, *20*, 273–297. [CrossRef]
38. Jiménez, B.; Holmes, E.; Heude, C.; Tolson, R.F.; Harvey, N.; Lodge, S.L.; Chetwynd, A.J.; Cannet, C.; Fang, F.; Pearce, J.T.M.; et al. Quantitative Lipoprotein Subclass and Low Molecular Weight Metabolite Analysis in Human Serum and Plasma by 1H NMR Spectroscopy in a Multilaboratory Trial. *Anal. Chem.* **2018**, *90*, 11962–11971. [CrossRef]
39. Benjamini, Y.; Hochberg, Y. Controlling the False Discovery Rate: A Practical and Powerful Approach to Multiple Testing. *J. R. Stat. Soc. Ser. B* **1995**, *57*, 289–300. [CrossRef]
40. Zhang, X.; Zhao, X.-W.; Liu, D.-B.; Han, C.-Z.; Du, L.-L.; Jing, J.-X.; Wang, Y. Lipid Levels in Serum and Cancerous Tissues of Colorectal Cancer Patients. *World J. Gastroenterol.* **2014**, *20*, 8646–8652. [CrossRef] [PubMed]
41. Mayengbam, S.S.; Singh, A.; Pillai, A.D.; Bhat, M.K. Influence of Cholesterol on Cancer Progression and Therapy. *Transl. Oncol.* **2021**, *14*, 101043. [CrossRef] [PubMed]

 applied sciences

MDPI

Article

Untargeted ^{1}H-NMR Urine Metabolomic Analysis of Preterm Infants with Neonatal Sepsis

Panagiota D. Georgiopoulou [1], Styliani A. Chasapi [1], Irene Christopoulou [2], Anastasia Varvarigou [2,*] and Georgios A. Spyroulias [1,*]

[1] Department of Pharmacy, University of Patras, 26504 Patras, Greece; pennygeo5@gmail.com (P.D.G.); stella.chimic@gmail.com (S.A.C.)

[2] Department of Paediatrics, University of Patras Medical School, General University Hospital, 26500 Patras, Greece; christop_irini@yahoo.gr

* Correspondence: varvarigou@upatras.gr (A.V.); g.a.spyroulias@upatras.gr (G.A.S.)

Abstract: One of the most critical medical conditions occurring after preterm birth is neonatal sepsis, a systemic infection with high rates of morbidity and mortality, chiefly amongst neonates hospitalized in Neonatal Intensive Care Units (NICU). Neonatal sepsis is categorized as early-onset sepsis (EOS) and late-onset sepsis (LOS) regarding the time of the disease onset. The accurate early diagnosis or prognosis have hurdles to overcome, since there are not specific clinical signs or laboratory tests. Herein, a need for biomarkers presents, with the goals of aiding accurate medical treatment, reducing the clinical severity of symptoms and the hospitalization time. Through nuclear magnetic resonance (NMR) based metabolomics, we aim to investigate the urine metabolomic profile of septic neonates and reveal those metabolites which could be indicative for an initial discrimination between the diseased and the healthy ones. Multivariate and univariate statistical analysis between NMR spectroscopic data of urine samples from neonates that developed EOS, LOS, and a healthy control group revealed a discriminate metabolic profile of septic newborns. Gluconate, myo-inositol, betaine, taurine, lactose, glucose, creatinine and hippurate were the metabolites highlighted as significant in most comparisons.

Keywords: metabolomics; NMR spectroscopy; neonatal sepsis; EOS; LOS; preterm birth

Citation: Georgiopoulou, P.D.; Chasapi, S.A.; Christopoulou, I.; Varvarigou, A.; Spyroulias, G.A. Untargeted ^{1}H-NMR Urine Metabolomic Analysis of Preterm Infants with Neonatal Sepsis. *Appl. Sci.* **2022**, *12*, 1932. https://doi.org/10.3390/app12041932

Academic Editor: Chiara Cavaliere

Received: 24 December 2021
Accepted: 8 February 2022
Published: 12 February 2022

1. Introduction

Neonatal sepsis is a systemic inflammatory response to infection, with a wide range and severity of symptoms [1]. The incidence varies all over the world and is quite different among high-, low-, and middle-income countries [2]. According to World Health Organization (WHO), 1.3 to 3.9 million cases are reported annually and 400,000 to 700,000 deaths worldwide [3]. Despite the progress in medical knowledge, neonatal care and antibiotic treatment, sepsis is considered one of the main reasons of morbidity and mortality, especially in very low birth weight (VLBW, <1500 g birth weight) and preterm infants (born before 37 weeks of pregnancy) [4–6]. Extremely preterm infants, whose gestational age (GA) is less than 28 weeks, are at greater risk of sepsis diagnosis [7,8]. However, several studies suggest that also late preterms are highly susceptible to developing sepsis [9–11].

Neonatal sepsis is classified as early onset sepsis (EOS), occurring within 72 h after birth, and late onset sepsis (LOS), 72 h after birth according to the onset time of the findings [12]. The incidence of EOS is 1 to 5 per 1000 live births, decreasing with intrapartum antibiotic therapy [13,14]. Associated with the increasing survival rate of preterm and VLBW infants, LOS rate presents an increase, with neonates weighing less than 750 g having the highest rate of LOS diagnosis [15,16].

Prematurity and low birth weight are among the main risk factors causing sepsis, with premature neonates having three to ten times higher possibility of sepsis diagnosis than normal weight full-term ones [1]. Risk factors and mortality rates differ among early-

and late-onset sepsis [17]. Among the EOS risk factors are fetal distress, low APGAR score, resuscitation of the neonate and multiple pregnancies. Additionally, for low-income countries, as EOS seems to be enhanced from factors such as inadequate antenatal care, unhealthy birth and late recognition of conditions that might induce infections to the mother or the neonate [18]. Factors that can increase risk of LOS are considered mainly invasive procedures, such as surgical interventions, intubation, mechanical ventilation, catheter/probe insertion, long-term parenteral nutrition, frequent blood sampling, low stomach acid and/or insufficient breastfeeding [1].

The pathogenesis of EOS is taking place during the intrapartum period, during which the responsible pathogens are transmitted from mother to neonate [19]. Pathogens causing EOS are usually colonized in maternal genitourinary tract and with the amniotic membrane rupture are transmitted to the fetus or during the labor to neonate [20]. Most frequent EOS's pathogens are the Gram (+) group B Streptococcus (GBS) followed by the Gram (−) *E. Coli* bacteria [21]. Pathogens causing LOS can be transmitted during labor or from the environment. LOS is usually caused by nosocomial or environmental pathogens, specifically Coagulase-negative staphylococci (CONS), Gram (−) bacilli and Fungi, especially *C. albicans*. Apart from differences in the pathogenesis and the time of onset, EOS and LOS have different clinical manifestations [22].

A combination of clinical signs and laboratory findings constitute the diagnosis procedure, with blood culture considered as the "gold" standard [23]. For the physicians, the diagnosis of sepsis is a challenge. The limited diagnostic accuracy of common laboratory tests, such as white blood cell indices and acute phase reactants contribute poorly to the early diagnosis of neonatal sepsis. The diagnostics' accuracy limitations, in combination with the neonate's prematurity and survival status as well as the ambiguity of early clinical signs, urge neonatologists to shut out sepsis [24]. Hence, the discovery of new biomarkers, which can easily be detected at an early stage, has occupied the medical and scientific community, with many studies already having been carried out [25].

In this scope, metabolomic analysis could take upon a fundamental role. Detecting and quantifying a wide variety of small molecules, intermediate or final components of metabolic pathways, metabolomics aim to identify the alterations caused by the condition of biological and medical interest. As the outcome of biochemical procedures regulated by proteins derived from genes expression, metabolomics provides the closest relation with the phenotype of an organism, at a specified time frame in correlation with endogenous and exogenous influences. The utilization of metabolomics as a tool for the validation of new biomarkers for early diagnosis or prognosis of pathophysiological conditions is constantly growing.

Before any metabolomic analysis, based on the main question to be answered, the process of the analysis has to be determined. A targeted metabolomics approach is selected when the aim is to measure the levels of a particular set of metabolites which are suspected to have a relation with a condition. An untargeted approach aims to detect and assign as many peaks as possible related to metabolites biofluid, tissue or cell extract under study, identifying the metabolic "fingerprint" [26]. Nuclear magnetic resonance (NMR) spectroscopy and mass spectrometry (MS) are the two dominant analytical techniques for metabolomic analysis of biofluids. NMR-based analysis offers numerous advantages and could combine chemometrics and basic clinical research [27].

In the field of neonatology, the interest for a prognostic and diagnostic tool for neonatal sepsis is continuously growing. A urine sample, reflecting a holistic view of the metabolism, is considered as the most appropriate biological fluid for analysis of newborns' metabolism via NMR spectroscopy. The impact of neonatal sepsis on newborn metabolism has been in the center of interest the last few years and application of metabolomic studies towards the investigation of a specific profile of septic newborns is on demand, but, to establish metabolites as biomarkers, further analysis must be conducted [28–30]. The aim of our study is to reveal those metabolites with differentiated levels among the septic neonates hospitalized in the Neonatal Intensive Care Unit (NICU) and control/healthy neonates.

Such metabolic alterations could lead to early diagnosis of EOS and LOS. To our knowledge, both medical treatment and parenteral nutrition of NICU hospitalization may induce essential metabolic drifts. For this reason, we chose to investigate the comparison of the sepsis metabolic profile with the metabolic profile of NICU hospitalized neonates without severe comorbidities except prematurity, complementary to healthy neonates, whose care after birth was taken over by their mother.

2. Materials and Methods

2.1. Study Population

The current study was conducted in the Neonatal Ward and the NICU of the General University Hospital, and the Department of Pharmacy at the University of Patras. Seventy-one (n = 71) neonates were recruited in the study, only after informed consent and parental permission having been provided. The neonates were separated into the following four groups: Group A: Septic neonates diagnosed with EOS (n = 23); Group B: Septic neonates diagnosed with LOS (n = 11); Group C: Preterm neonates without any clinical sign or symptom of sepsis or other serious morbidity (n = 14) hospitalized in the NICU; Group D: Healthy preterm neonates (n = 23) not separated from their mother after birth. All NICU neonates were premature, with GA ranging from 25 to 36 GA weeks, while all healthy non NICU preterm neonates ranged from 35 to 36 GA weeks. For each neonate, the following were recorded: (a) perinatal data such as gestational age (GA), birth weight (BW), sex, delivery mode, premature rupture of membranes and Apgar score; (b) clinical data such as mechanical ventilation, treatment with antibiotics, positive blood cultures, C reactive protein (CRP), and breast feeding. All clinical characteristics and laboratory findings of the neonates participated in this study are listed in Table 1. The biological fluid under study was urine and the samples were collected in the first 24 h after birth for EOS neonates and in the third day of the extrauterine life for LOS neonates. The collection was carried out with the use of adhesive pediatric urine collection bags and 1.5 mL of each collected urine sample was transferred to sterile vials and remained at −80 °C until the analysis.

Table 1. Clinical characteristics and laboratory findings for septic and control neonates. Median and (minimum–maximum) values of the GA, BW, Apgar Score for the first and tenth minute of life, the number (percentage) of the male sex, the delivery mode (cesarian section), the small for gestational age (SGA) infants, the newborns with premature rupture of membranes (>18 h), treatment with mechanical ventilation and/or antibiotics, blood culture positive (Gram-positive, Gram-negative and fungi) and c reactive protein (CRP).

	Group A EOS (n = 23)	Group B LOS (n = 11)	Group C Control NICU (n = 14)	Group D Control Non NICU (n = 23)
Male sex (n, %)	8 (35)	5 (45)	9 (64)	18 (75)
GA (weeks)	34 (26–36)	34 (25–36)	35 (31–36)	36 (35–36)
BW (gr)	2150 (770–4060)	1820 (690–2900)	2085 (1630–3540)	2740 (2100–3700)
Small for GA (n, %)	5 (22)	4 (36)	2 (14)	0(0)
Cesarian Section (n, %)	8 (73)	8 (73)	11 (79)	13 (57)
Apgar Score 1st min	8 (3–9)	8 (3–9)	8 (7–9)	9 (5–9)
Apgar Score 10th min	9 (8–10)	9 (8–10)	9.5 (8–10)	9 (7–10)
Antibiotics (n, %)	23 (100)	11 (100)	1 (7)	11 (48)
Premature Rupture of Membranes >18 h (n, %)	1 (4.3)	0 (0)	3 (21.4)	0 (0)
Mechanical Ventilation (n, %)	17 (71)	7 (64)	4 (14)	0 (0)
Nutrition	No	No	No	Breast milk
Laboratory findings				
B.C negative, CRP positive findings (n, %)	17 (74)	5 (45)	0 (0)	0 (0)
B.C positive, gram (+) (n, %)	4 (17)	5 (45)	0 (0)	0 (0)
B.C positive, gram (−) (n, %)	1 (4)	1 (10)	0 (0)	0 (0)
B.C positive, Fungi (n, %)	1 (4)	0 (0)	0 (0)	0 (0)

2.2. Ethical Statement

The study was approved by the General University Hospital of Patras Human Research Ethics Committee and all the procedures performed in this study involving human participants were in accordance with the ethical standards of the institutional and/or national research committee and with the 1964 Helsinki declaration and its later amendments or comparable ethical standards. A written informed consent was obtained from parents.

2.3. NMR Sample Preparation

The urine samples were stored at −80 °C until analysis. For NMR analysis, frozen urine samples were thawed at room temperature and centrifuged at 12,000 rpm for 10 min at 4 °C. For each sample, 540 μL of the supernatant was mixed with 60 μL potassium phosphate buffer (1.5 M KH_2PO_4 in H_2O containing 4% 2 Mm NaN_3 and 10 mM 4,4-dimethyl-4-silapentane-1-sulfonic acid (DSS), pH 4.2). The mixture was vortexed and 590 μL of it was transferred into a 5 mm NMR tube (Bruker BioSpin srl).

2.4. NMR Experiments

The ^{1}H-NMR spectra were recorded at 298 K on a Bruker Avance III HD 700 MHz NMR spectrometer equipped with a cryogenically cooled 5.0 mm ^{1}H/^{13}C/^{15}N/D Z-gradient probe. Two kinds of ^{1}H-NMR spectra were recorded for each urine sample of the study's participants. A mono-dimensional (1D) NMR spectrum was acquired using a standard NOESY (noesygppr1d.comp; Bruker BioSpin, Billerica, MA, USA) pulse sequence for water suppression, to reveal all the detectable ^{1}H signals of metabolites. Due to the complexity of urine NMR spectra, as they consist of numerous peaks, presented in a non-distinctive manner, and sometimes overlapped, the identification and further quantification of metabolites using only 1D NMR spectra cannot be accomplished [31]. Hence, for each urine sample, a two-dimensional (2D) *J*-resolved (jresgpprqf.comp; Bruker BioSpin) spectrum acquired, separating the chemical shifts and J-couplings into two different dimensions, making it a very useful NMR experiment for metabolite assignment in metabolomics [32]. Accurately, the acquisition parameters for the ^{1}H 1D NMR and 2D *J*-resolved spectra were 64 scans, 4 dummy scans, FID size 64.536, a spectral width of 10,504 Hz, 3.1 s acquisition time, 2 s a relaxation delay, and 100 ms mixing time [33].

2.5. Data Processing

After the NMR data acquisition, all spectra were manually processed for phase and baseline corrections and calibrated to the internal standard's DSS peak at 0.00 ppm. The processing was performed using the TopSpin 4.1.1 (Bruker BioSpin srl). For further statistical analysis, all ^{1}H 1D NOESY NMR spectra were transformed into binned numerical data. Each spectrum was fragmented into buckets of 0.02 ppm width for the spectral region 0.70–9.50 ppm, using the AMIX software (Bruker BioSpin). The spectral areas of 4.50–6.00 ppm, 2.70–2.80 ppm and 1.75–1.77 ppm where ^{1}H signals of water, urea and internal standard DSS are observed, were excluded. Chenomx software (NMR Suite Version 9.0, Edmonton, AB, Canada), the online Human Metabolome Database (HMDB) and data from literature, were utilized for the assignment of important and discriminant proton NMR signals of urine samples.

2.6. Statistical Analysis of NMR Data

The statistical analysis of the spectral binning data was performed using the online tool MetaboAnalyst 5.0 [34]. Multivariate analysis (MVA), consisting of the unsupervised principal component analysis (PCA) and supervised partial least squares discriminant analysis (PLS-DA), was applied on the ^{1}H NMR data after Pareto normalization. This normalization was selected, since NMR information does not deviate much from the original compared to autoscaling, and fewer errors related to noisy spectral regions without biological impact are conducted [35]. PCA and PLS-DA plots were examined to extract information about group clustering and potential outliers. Loadings values from PCA model and variable importance in projection (VIP) scores >1 from PLS-DA were the main evidence source about metabolites with differentiation between groups, responsible for the group clustering. The quality of PLS-DA was ensured using the parameters as follows: goodness of fit R^2 and the goodness of predictability Q^2, after 10-fold cross-validation test. For statistical correlation of NMR data, a mono-dimensional (1D) statistical total correlation spectroscopy analysis was performed using the muma R package [36], where pareto scaled data were analyzed and the represented pseudo-NMR spectrum displayed the covariance (height) and the Pearsons correlation coefficient (color) of all spectral variables (buckets) with the variable of interest being the "driver peak" or "driver signal". Additionally, univariate analysis was performed on the successfully identified metabolites, targeting only the non-overlapping metabolites' peaks for integration. Non-parametric Kruskal–Wallis test, followed by a false discovery rate (FDR) correction, was conducted using RStudio and metabolites with p-value <0.05 were characterized as significant.

2.7. Pathway Analysis and Visualization

A pathway analysis module of MetaboAnalyst 5.0, integrating enrichment analysis and pathway topology analysis through a Google Maps-style visualization system, was adopted to identify the metabolic pathways associated with the statistically significant metabolites for EOS, LOS and healthy preterms. After the import of metabolites' compound names as data input, the Homo Sapiens pathway library from KEGG was selected. A hypergeometric test and relative betweenness centrality were preferred for over-presentation and topological analysis, respectively.

3. Results

A total of 45 metabolites were successfully detected and assigned in urine samples of the preterm neonates. Due to the complexity and peak overlapping of the ^{1}H NMR urine spectra, statistically important and peaks with discriminant signals were recorded for the qualitative identification of the neonate's urine metabolism. Table S1 represents the list of all the assigned metabolites (Supplementary materials Table S1) and the main pathway they may belong to. Further statistical analysis of the binned data was based on the reported chemical shifts to associate statistical important spectral regions with specific metabolites.

3.1. Metabolic Profile Alterations of Preterms with EOS

3.1.1. EOS Preterms Versus Non NICU Healthy Preterms

Twenty-three preterm infants (n = 23), hospitalized in NICU, were diagnosed with EOS and their urine NMR metabolic profiles were compared to the first day of life spectroscopic data of the control group, consisting of twenty (n = 20) healthy preterm neonates, treated by their mothers immediately after birth and non-hospitalized in NICU (non NICU preterms). The generated data, consisting of 359 spectral bins, were processed using Pareto normalization. The numerical data matrix was applied as input for the unsupervised PCA and the supervised PLS-DA. Both showed a clear discrimination between the two groups. According to PC1 and PC2, explaining the 73.4% of the total variance, loadings belonging to an unknown pattern of multiple peaks between 7.40–7.50 ppm and at 1.40 ppm, observed to the majority of the NMR spectra of the NICU urine samples (20 out of 23 spectra of EOS samples), were responsible for the separation (Supplementary materials Figure S1). Additionally, gluconate and lactose differed between the two groups, with EOS neonates having the higher concentration level of gluconate and lower of lactose. The PLS-DA model (Figure 1a) shows a discrimination between the two groups, with an R^2 of 0.664 and Q^2 of 0.409 for the third component after a 10-fold cross-validation test, supporting the unsupervised PCA's clustering. VIP scores, with colored boxes on the right representing the differentiated relative concentrations of the corresponding buckets in each group (blue—low—and red—high—relative concentration) were in accordance with loadings (Figure 1b) and those with higher scores reveal lower intensity of buckets related to ppm of taurine, myo-inositol and betaine, creatinine for EOS and higher intensity of the unknown area 7.40–7.50 ppm. The box plots of the metabolites responsible for the group clustering confirm the concentration's alteration among the compared groups (Figure 1c). The Y axis represents the normalized concentration of the corresponding spectral region. The obtained negative values come as a result of the total spectral area normalization. The unassigned spectral region of two different types of peaks that resonate at 1.40 ppm statistically correlate with the multiple peaks at the aromatic region of 7.40–7.50 ppm. The statistical correlation was confirmed with the 1D-STOCSY (Supplementary materials Figure S2) that indicates all the correlations of all spectral variables (buckets). To further examine and highlight the structural correlations with the spectral variables that resonate at 7.40–7.50 ppm, we set as "driver peak" the variable x.1.41, which includes the signals at 1.40 ppm and the observed maximum correlation displayed between the variable x.1.41 and x.7.51. Thus, they may belong to the same metabolite or group of metabolites derived from medication or parenteral nutrition. We attribute the unknown spectral pattern at 7.40–7.50 ppm and at 1.40 ppm as a NICU metabolic-induced characteristic, since it is present predominantly in the ^{1}H NMR spectra from NICU urine samples (Figure 2). Most of the preterm neonates hospitalized in NICU are under antibiotic treatment because of sepsis or of other infectious condition suspicion; also, prophylactic administration of broad-spectrum antibiotics is unfortunately very common. Antibiotics can lead to adverse effects, including necrotizing enterocolitis (NEC) and LOS [37]. Patton et al. studied their impact on fecal metabolome of preterm infants, but urine metabolome has not been investigated yet [38]. Our study suggests the further analysis of these unidentified spectral regions as antibiotic outcome on ^{1}H NMR spectra of urine samples. However, in depth investigation of the antibiotic and/or additional medication effect on the urine NMR metabolic profile is beyond the scope of this research.

3.1.2. EOS Preterms versus NICU Control Preterms

^{1}H NMR urine metabolic profile of the first day of life from twenty-three (n = 23) preterm infants with EOS was compared to that of thirteen (n = 13) preterms hospitalized in NICU without any sign or symptom of sepsis or other infection. This comparison was conducted complementary to the group of preterms without the need of hospitalization, to examine and reduce the effect of NICU hospitalization on newborns urinary metabolism. Multivariate analysis of EOS and NICU control preterms displayed less distinctive classification (Supplementary Materials Figure S3a) regarding the comparison with healthy non-NICU preterms (Supplementary Materials Figure S1a). The PLS-DA model (Figure 3a) separated the two groups, but the low value of the R^2 = 0.491 and the negative Q^2 indicate an overfitted model with low predictability. The buckets responsible for this group separation, as resulting from the PCA loadings plot (Supplementary Materials Figure S3b) and from VIP scores, are related to gluconate, threonine/lactate and 7.40 ppm chemicals shifts of unknown peaks (Figure 3b). Gluconate seems to be present in higher concentration on EOS group, but some buckets related to gluconate (4.09 ppm, 4.15 ppm) have higher VIP score in the control group. This may be justified by the shift of gluconate peaks in different spectra or the presence of sugars, such as glucose and lactose. The pattern of 7.40, 1.40 ppm had higher intensity for control neonates, and may be related to different medical treatment and nutrition. In accordance with prior analysis, EOS neonates differ from controls at the 3.25 ppm spectral region, where mainly betaine is located, overlapping myo-inositol and taurine.

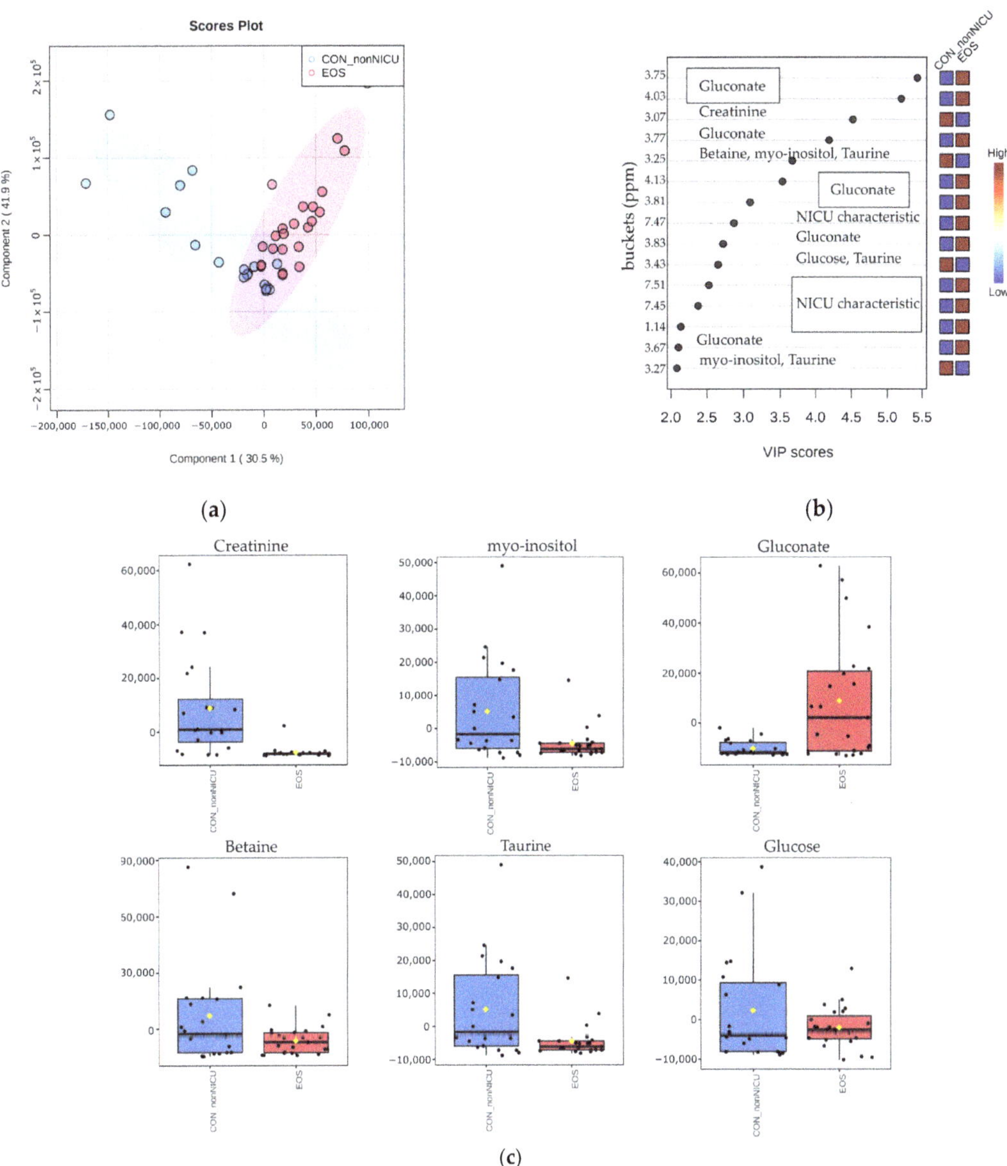

Figure 1. Multivariate analysis of NMR data belonging to neonates diagnosed with EOS (pink circles) and healthy neonates without need for NICU hospitalization (blue circles). (**a**) PLS-DA scores plot of the EOS and healthy non-NICU preterms. (**b**) VIP scores and metabolites related to buckets with different concentration among the two groups. (**c**) Box plots of normalized concentration for the discriminant metabolites.

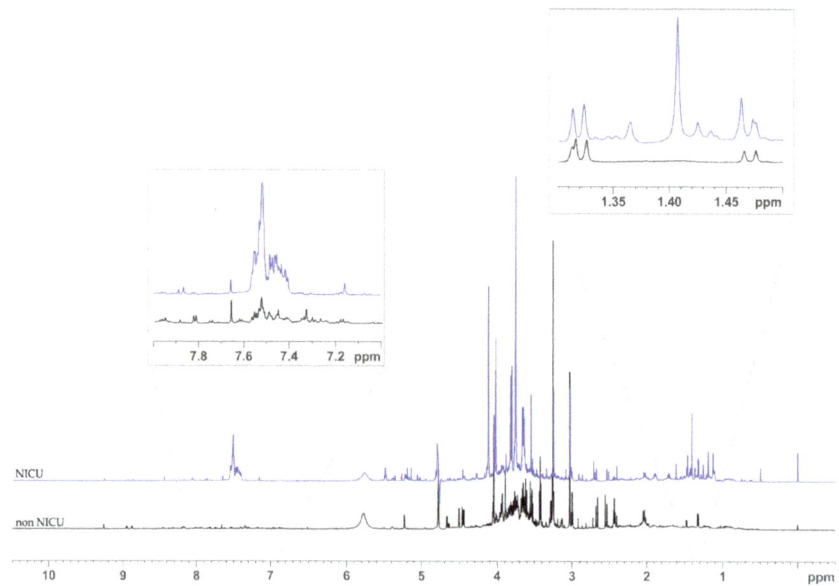

Figure 2. Spectral regions of unassigned chemical shifts mostly present on ^{1}H NMR spectra of NICU samples (blue spectrum) and healthy non-NICU preterms (black spectrum).

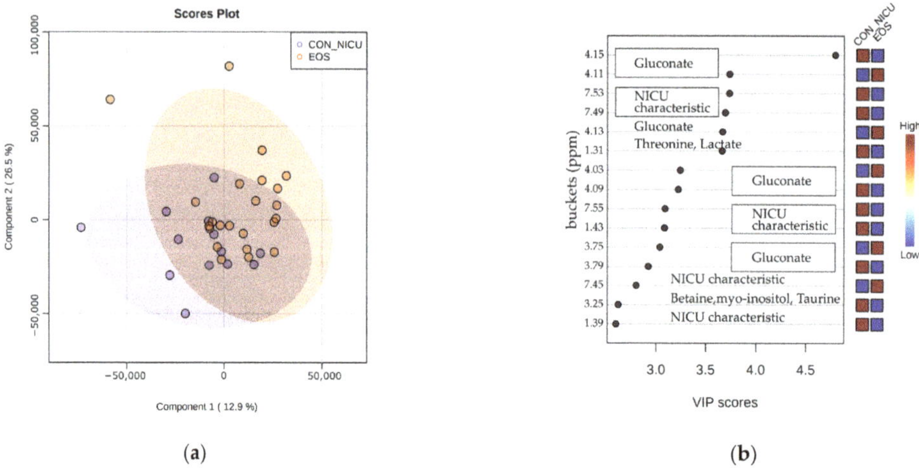

Figure 3. Multivariate analysis of NMR data belonging to neonates diagnosed with EOS (orange circles) and neonates hospitalized in NICU without EOS (purple circles). (**a**) PLS-DA scores plot of the EOS and control group. (**b**) VIP scores and metabolites related to buckets with different concentration among the two groups.

3.1.3. EOS Metabolic Profile Progression between First and Third Day of Life

With the perspective for the validation of a metabolite for further analysis or the identification of a group of metabolites characteristic of the septic urine metabolome, the progression of the metabolome throughout the time of the condition needs to be examined. In our study a comparative analysis of urine metabolomes between the first and the third day of septic newborns' extrauterine life was conducted. For ten (n = 10) out of the twenty-three neonates diagnosed with EOS, urine samples and NMR data from the third day of neonates' life are not available due to handling inaccuracies during sample collection or other excluding criteria regarding the NMR spectra (e.g., baseline and phase distortions) and the presence of lipids, characterized of broad peaks overlapping peaks of small molecules studied in this analysis. Hence, a total of twenty-six (n = 26; paired 1st and 3rd day of life) urine ^{1}H NMR spectra, corresponding to each one of the thirteen (n = 13) EOS preterms, were analyzed. An initial classification of the two groups was performed via PCA, where the third principal component, explains the 79.9% of the cumulated variance (Supplementary materials Figure S4). PLS-DA corroborated and strengthened the clustering, providing a valid and reliable model (Figure 4a). The cross-validation indicated the use of the first two components as optimal for building the classification model, with $R^2 = 0.613$ and $Q^2 = 0.296$. VIP scores were in accordance with the PCA's loadings and indicated reduced concentration of myo-inositol, gluconate, betaine, taurine, and creatinine on the third day (Figure 4b). The normalized concentration's differentiated levels of the reported metabolites are clearly represented on their associated box plots (Figure 4c). The alteration of myo-inositol levels between the first and the third day of life has been previously reported for healthy full-term (>38 GA) neonates [33]. Specifically, in association with sepsis, increased levels of myo-inositol have also been detected on the work of Sarafidis et al. for LOS at the day of the disease's onset [28]. The decrease in creatinine is in accordance with the Fanos et al. findings for septic neonates [30].

3.2. *Metabolic Profile Alterations of Preterms with LOS*

3.2.1. LOS Preterms versus Non NICU Healthy Preterms

Spectroscopic data of urine samples collected on the third day of life from eleven (n = 11) preterm neonates that developed LOS were compared to the third day's ^{1}H NMR urine profile of twelve (n = 12) non-NICU healthy preterms. The unsupervised PCA, explaining the 75.1% of the total variability within the first three components (PC1 = 43.3%, PC2 = 18.2% and PC3 = 13.6%) indicated a clear tendency for clustering the two groups. Loadings with the greatest impact caused this form of PCA plot, agreed with EOS results and reinforced the claim of the great impact of hospitalization in NICU on urine metabolome (Supplementary materials Figure S5). The supervised PLS-DA, after the cross-validation resampling method, present a reliable and predictable model with $R^2 = 0.861$ and $Q^2 = 0.788$ for the second component (Figure 5a). As expected, VIP scores highlighted gluconate and the pattern of 1.40, 7.40–7.50 ppm (Figure 5b), with significant alterations of normalized concentration among groups, clearly displayed on the box plots (Figure 5c). Multivariate analysis of LOS metabolic profile reveals similar metabolic alterations with EOS. This analysis did not add something different compared to the analysis about EOS and healthy non-NICU preterms, and it is primary evidence that the septic profile does not dramatically change in relation to time within the first three days of neonate's life.

3.2.2. LOS Preterms Versus NICU Control Preterms

Following the same procedure with the comparisons for EOS neonates, in order to eliminate the impact of hospitalization and dietary or drugs urine excreted metabolites in urine LOS data compared to the NICU control group. In total, eleven (n = 11) ^{1}H NMR urine LOS profiles were compared with nine (n = 9) ^{1}H NMR urine NICU control profiles from the third day of life. Scores plot of the PCA model (Supplementary materials Figure S5a) primarily did not reveal strong discrimination and resembles the PCA plot clustering of EOS analysis. Additively, presents different urine metabolic profile with the same elevations of myo-inositol, betaine, gluconate, taurine and NICU characteristics (Supplementary materials Figure S6b). The PLS-DA model is characterized by low capacity of predictability with negative Q^2 values (Figure 6a). Beyond these metabolites, VIP scores revealed higher concentration of buckets belonging to sugars, for the control group (Figure 6b). The bucket of 1.17 ppm belongs to a crowded ^{1}H NMR region, without a specific metabolite present to the total of urine samples.

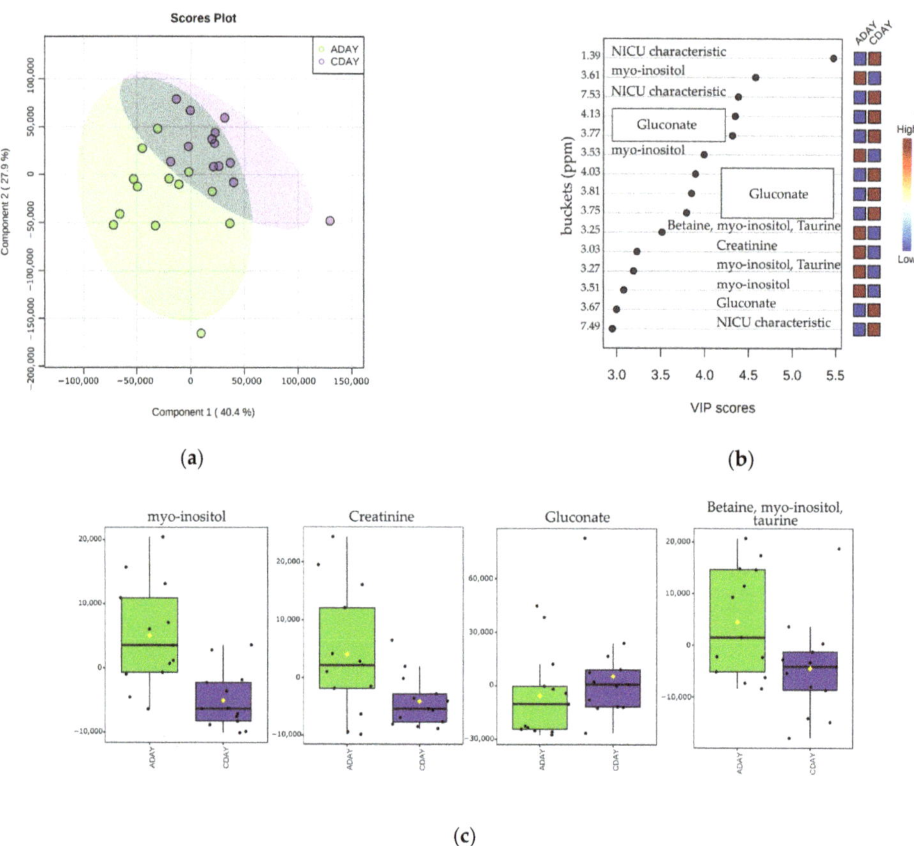

Figure 4. Multivariate analysis of NMR data belonging to urine samples of neonates diagnosed with EOS the first (green circles) and the third day (blue circles) of their life. (**a**) PLS-DA scores plot of the first- and third-day's samples. (**b**) VIP scores and metabolites related to buckets with different concentration among the two groups. (**c**) Box plots of normalized concentration for the discriminant metabolites.

3.3. Metabolic Profile Alterations between EOS and LOS Neonates

LOS and EOS neonates, except the onset of the disease, present different clinical symptoms and vary on the severity of the outcome. This differentiation can be reflected also on the metabolism. So, additively to separate comparisons between EOS, LOS and control groups, multivariate analysis was conducted between the urine metabolic profile of the first day of life for twenty-three (n = 23) EOS neonates and third day of life for eleven (n = 11) LOS neonates. For the PCA model (Supplementary materials Figure S7), until the PC3 the variance explained was 70.1% and the plot showed a tendency of clustering the urine metabolic profiles of each group, which became clear on the PLS-DA plot (Figure 7a). The metabolic alterations, according to VIP scores (Figure 7b) and box plots (Figure 7c), resemble the differences among the first and third day of EOS. So, the correlation between EOS and LOS cannot be validated as this metabolic outcome may reflect the adaptation of the neonate to the extrauterine life. To shed light on the alterations of the septic metabolome over the days, targeted metabolomic analysis of specific metabolites already known or suspected based on metabolomics results for their association with sepsis would offer a more certain approach.

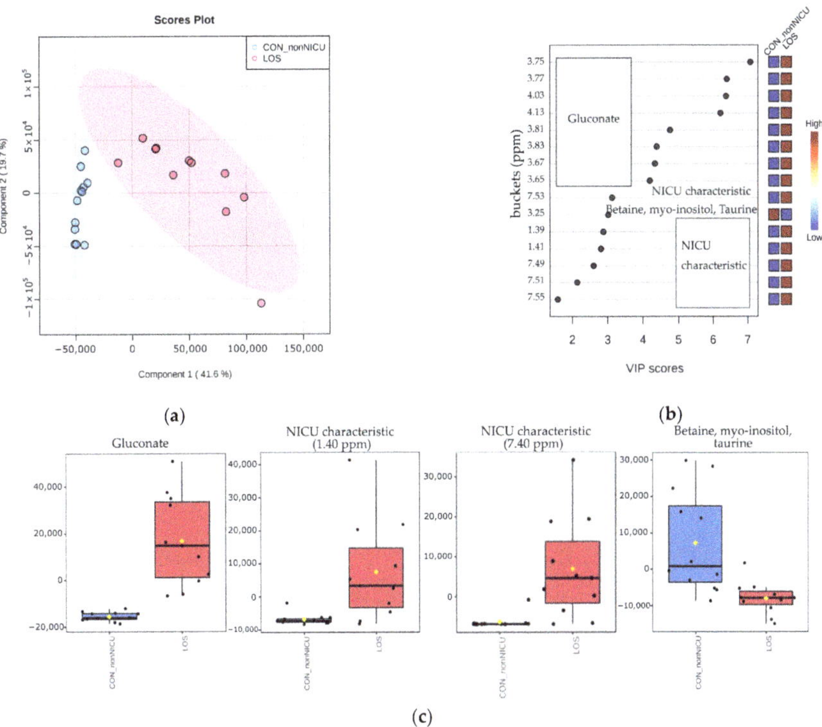

Figure 5. Multivariate analysis of NMR data belonging to neonates diagnosed with LOS (pink circles) and control neonates without need for hospitalization (blue circles). (**a**) PLS-DA scores plot of the LOS and healthy non-NICU preterms without need for hospitalization. (**b**) VIP scores and metabolites related to buckets with different concentration among the two groups. (**c**) Box plots of normalized concentration for the discriminant metabolites.

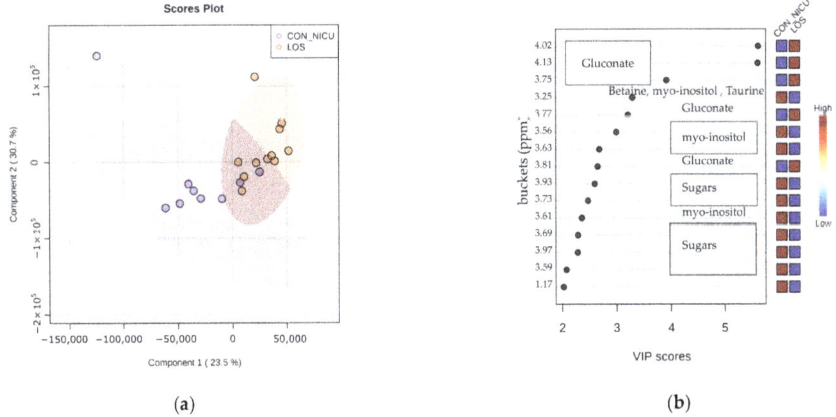

Figure 6. Multivariate analysis of NMR data belonging to neonates diagnosed with LOS (orange circles) and control neonates of NICU (blue circles). (**a**) PLS-DA scores plot of the LOS and control. (**b**) VIP scores and metabolites related to buckets with different concentration among the two groups.

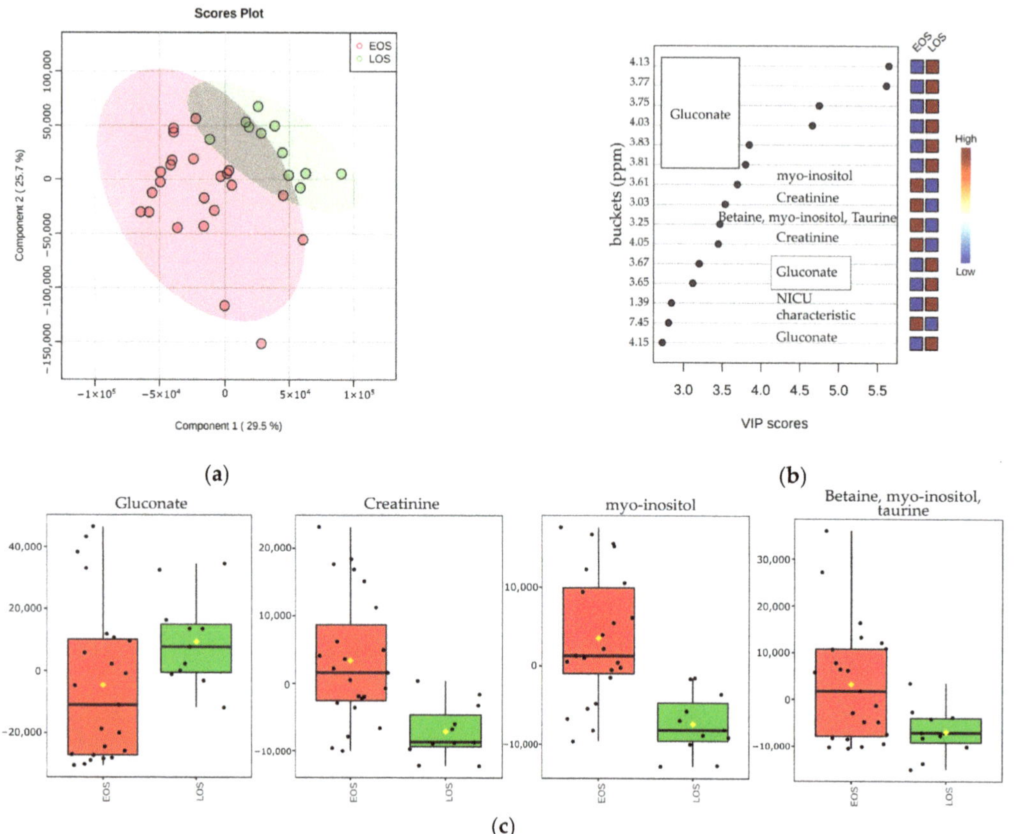

Figure 7. Multivariate analysis of NMR data belonging to urine samples of neonates diagnosed with EOS (red circles) and neonates with LOS (green circles). (**a**) PLS-DA scores plot of EOS and LOS samples. (**b**) VIP scores and metabolites related to buckets with different concentration among the two groups. (**c**) Box plots of normalized concentration for the discriminant metabolites.

3.4. Univariate Statistical Analysis

A univariate statistical analysis was performed on specific metabolites with discriminant peaks. Metabolites with *p*-value < 0.05 (Table 2) were characterized as statistically significant. Between septic and healthy non-NICU preterms, additively to multivariate results, lactose and hippurate were highlighted as significant for EOS, and dimethylglycine for LOS group. For both septic groups decreased levels of taurine, betaine and increased levels of gluconate also shown by multivariate analysis were reinforced by univariate analysis. EOS and LOS metabolites' comparison confirmed the initial results obtained through multivariate analysis regarding myo-inositol's differentiation. Septic groups and control NICU preterms did not highlight any statistically significant metabolite. The box plots of the statistically significant metabolites represent the differentiation of their relative intensity between septic and healthy control non-NICU neonates (Supplementary materials Figure S8).

Table 2. Statistically significant metabolites and their *p*-value for EOS and LOS groups.

	Metabolites	*p*-Value	Septic Group
EOS versus healthy non NICU preterms	Taurine	0.004	↓
	Gluconate	8×10^{-5}	↑
	Lactose	1×10^{-4}	↓
	Hippurate	7×10^{-5}	↓
LOS versus healthy non NICU preterms	Gluconate	0.0007	↑
	Lactose	0.01	↓
	Betaine	0.006	↓
	N, *N*-Dimethylglycine	0.005	↓
	Hippurate	0.03	↓

3.5. Pathway Analysis

The identified significant metabolites from multivariate and univariate analysis were implemented into MetaboAnalyst pathway analysis module to determine a qualitative aspect of all the possibly affected metabolic pathways. A separate analysis for the EOS and LOS group was selected, as except myo-inositol, betaine, gluconate, taurine, lactose, creatinine and hippurate which were statistically significant for both groups, glucose and *N*, *N*-Dimethylglycine were characterized as significant only for EOS and LOS group, respectively. A metabolic pathway analysis depicted ten (n = 10) altered metabolic pathways for the EOS (Figure 8a, Supplementary materials Table S2) and nine (n = 9) for the LOS group (Figure 8b, Supplementary materials Table S3). The representation of all the identified pathways was based on the pathway impact (*x*-axis) and the calculated *p*-value (*y*-axis). Among the nine metabolic pathways for EOS, ascorbate/ alderate metabolism and taurine/hypotaurine metabolism were the most significant, with p-value < 0.05. Taurine/hypotaurine metabolism pathway had the largest impact factor (0.42), followed by inositol phosphate pathway (0.13). Myo-inositol and taurine were the metabolites involved in these pathways (Supplementary materials Table S2). Regarding the LOS, significant metabolites, glycine/serine, and threonine metabolism pathway, where betaine and *N*,*N*-Dimethylglycine are involved, had the lowest *p*-value (0.01). The pathway impact scores did not present large differences from EOS pathway analysis.

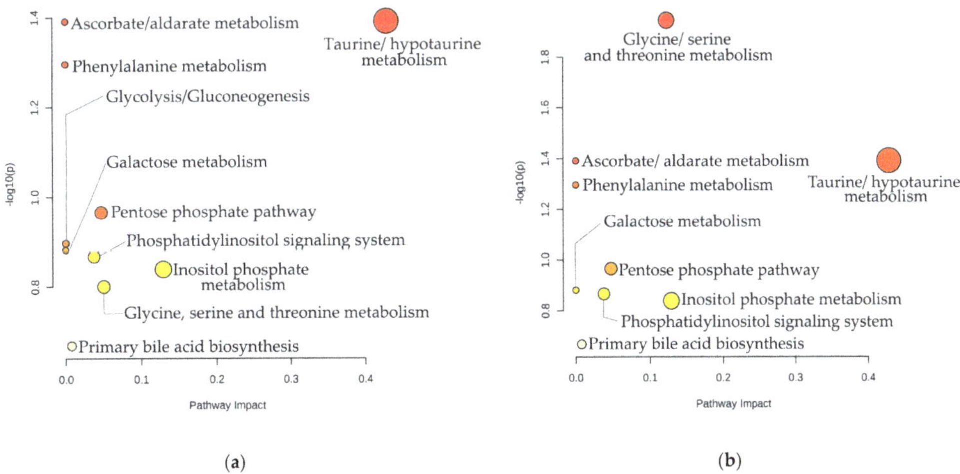

Figure 8. Graphically representation of the pathway analysis. Each cycle represents a metabolic pathway, while the color and the size are based on *p*-value and pathway impact, respectively. (**a**) Pathway analysis of EOS group significant metabolites. (**b**) Pathway analysis of LOS significant metabolites.

4. Conclusions

The findings of our study indicate a discrete metabolic profile of septic neonates. NMR based metabolomic approach revealed the relation among septic and control neonates, highlighting gluconate, myo-inositol, hippurate, taurine, *N*, *N*-Dimethylglycine, betaine, creatinine, glucose and lactose as significant metabolites. Our study reported, for the first time, altered urinary amounts of betaine in EOS and LOS neonates and *N*, *N*-Dimethylglycine in LOS neonates. Differentiated concentration levels of taurine and hippurate via LC-MS analysis have been previously reported by Sarafidis et al., and through UPLC-MS by Mardegan et al. [28,29]; however, our study was the first to detect them in urine samples of septic neonates via NMR. The utilization of the two control groups and their discrete analysis, based on the NICU hospitalization, showed that NICU treatment has a significant impact on neonates' urine metabolome. The observed spectral pattern indicative for the most of the NICU neonates, suggests that it is related to endogenous or exogenous metabolites of the personalized nutrition or medical treatment. Additionally, the impact of nutrition is confirmed from the greatly elevated levels of gluconate for the septic group and lactose for neonates fed from their mothers. Differentiation between EOS and LOS, and the adaptation of the fetus to neonate during the first days of extrauterine life that occur in parallel are reflected to the metabolism. Changes through the first days of life, associated with EOS and LOS, highlight the necessity for chronological coupled sampling with the onset and specific time of the disease progression. This research builds on the power of NMR metabolomic analysis to determine the status of an entire organism by a small amount of non-invasive collected biological sample. The establishment of NMR analysis of metabolome for clinical research in the field of neonatology, leading to large-scale multicenter studies, gives new and promising perspectives for its incorporation into the clinical daily routine and the validation of new combined diagnostic biomarkers.

Supplementary Materials: The following supporting information can be downloaded at: https://www.mdpi.com/article/10.3390/app12041932/s1; Figure S1: Scores and loadings plot of PCA for NMR data belonging to neonates diagnosed with EOS (pink circles) and healthy neonates without need for NICU hospitalization (blue circles). (a) PCA scores plot of the EOS and healthy non-NICU preterms. (b) Loadings plot of PC1 and PC2; Figure S2: 1D-STOCY pseudo-NMR spectrum of correlation coefficients to the other signals in the median urine NMR spectrum and maximum intensity correlation of peaks are color encoded and projected into statistical difference spectra: "driver peak" was set the one at 1.41 ppm; Figure S3: Scores and loadings plot of PCA for NMR data belonging to neonates diagnosed with EOS (orange circles) and neonates hospitalized in NICU without EOS (purple circles). (a) PCA scores plot of the EOS and control group. (b) Loadings plot of PC1 and PC2; Figure S4: Scores and loadings plot of PCA for NMR data belonging to urine samples of neonates diagnosed with EOS the first (green circles) and the third day (blue circles) of their life. (a) PCA scores plot of the first- and third-day's samples. (b) Loadings plot of PC1 and PC2; Figure S5: Scores and loadings plot of PCA for NMR data belonging to neonates diagnosed with LOS (pink circles) and control neonates with-out need for hospitalization (blue circles). (a) PCA scores plot of the LOS and healthy non-NICU preterms without need for hospitalization. (b) Loadings plot of PC1 and PC2; Figure S6: Scores and loadings plot of PCA for NMR data belonging to neonates diagnosed with LOS (orange circles) and control neonates of NICU (purple circles). (a) PCA scores plot of the LOS and control group. (b) Loadings plot of PC1 and PC2; Figure S7: Scores and loadings plot of PCA for NMR data belonging to urine samples of neonates diagnosed with EOS (red circles) and neonates with LOS (green circles). (a) PCA scores plot of EOS and LOS group. (b) Loadings plot of PC1 and PC2; Figure S8: Box plots of the statistically significant metabolites highlighted from univariate analysis with p-value < 0.05, between healthy control non-NICU neonates, LOS and EOS groups; Table S1: 1H NMR Chemical Shifts of Metabolites detected in urine samples of neonates and their main metabolic pathway; Table S2: Detailed results from the pathway analysis of the EOS group's significant metabolites; Table S3: Detailed results from the pathway analysis of the LOS group's significant metabolites.

Author Contributions: Study conception and design, I.C., A.V. and G.A.S.; patient enrolment and management: I.C. and A.V.; collection of clinical data and urine samples: I.C.; NMR analysis: S.A.C.; statistical analysis, biostatistics, and computational analysis, P.D.G.; results interpretation, P.D.G., S.A.C., I.C., A.V. and G.A.S.; writing—original draft preparation: P.D.G. and S.A.C.; writing—review and editing: P.D.G., S.A.C., I.C., A.V. and G.A.S.; supervision, A.V. and G.A.S. All authors have read and agreed to the published version of the manuscript.

Funding: The work was supported by the INSPIRED (MIS 5002550) which is implemented under the Action 'Reinforcement of the Research and Innovation Infrastructure,' funded by the Operational Program 'Competitiveness, Entrepreneurship and Innovation' (NSRF 2014–2020) and co-financed by Greece and the European Union (European Regional Development Fund). Additionally, EU FP7 REGPOT CT-2011-285950—"SEE-DRUG" project is acknowledged for the purchase of UPAT's 700 MHz NMR equipment.

Institutional Review Board Statement: This study was approved by the General University Hospital of Patras human research ethics committee (protocol code 7976/24.4.17, date of research committee approval: 217/ 31.3.2017 and study number approval: 154/16.3.2017 by the research ethics and deontology committee) all procedures performed in studies involving human participants were in accordance with the ethical standards of the institutional and/or national research committee and with the 1964 Helsinki declaration and its later amendments or comparable ethical standards.

Informed Consent Statement: Informed consent and parental permission were obtained from all the parents of the participants involved in the study.

Data Availability Statement: Data and R script are available from the corresponding authors upon reasonable request.

Conflicts of Interest: The authors declare that they have no conflict of interest.

References

1. Shane, A.L.; Sánchez, P.J.; Stoll, B.J. Neonatal Sepsis. *Lancet* **2017**, *390*, 1770–1780. [CrossRef]
2. Vergnano, S.; Sharland, M.; Kazembe, P.; Mwansambo, C.; Heath, P.T. Neonatal Sepsis: An International Perspective. *Arch. Dis. Child. Fetal Neonatal Ed.* **2005**, *90*, 220–224. [CrossRef]
3. World Health Organization (WHO). *Global Report on the Epidemiology and Burden of Sepsis: Current Evidence, Identifying Gaps and Future Directions*; World Health Organization: Geneva, Switzerland, 2020; ISBN 9789240010789.
4. Fanaroff, A.A.; Wright, L.L.; Stevenson, D.K.; Shankaran, S.; Donovan, E.P.; Ehrenkranz, R.A.; Younes, N.; Korones, S.B.; Stoll, B.J.; Tyson, J.E.; et al. Very-Low-Birth-Weight Outcomes of the National Institute of Child Health and Human Development Neonatal Research Network, May 1991 through December 1992. *Am. J. Obstet. Gynecol.* **1995**, *173*, 1423–1431. [CrossRef]
5. Shah, G.S.; Budhathoki, S.; Das, B.K.; Mandal, R.N. Risk Factors in Early Neonatal Sepsis. *Kathmandu Univ. Med. J. (KUMJ)* **2006**, *4*, 187–191.
6. Lim, W.H.; Lien, R.; Huang, Y.C.; Chiang, M.C.; Fu, R.H.; Chu, S.M.; Hsu, J.F.; Yang, P.H. Prevalence and Pathogen Distribution of Neonatal Sepsis among Very-Low-Birth-Weight Infants. *Pediatrics Neonatol.* **2012**, *53*, 228–234. [CrossRef]
7. Greenberg, R.G.; Kandefer, S.; Do, B.T.; Smith, P.B.; Stoll, B.J.; Bell, E.F.; Carlo, W.A.; Laptook, A.R.; Sánchez, P.J.; Shankaran, S.; et al. Late-Onset Sepsis in Extremely Premature Infants: 2000–2011. *Pediatr. Infect. Dis. J.* **2017**, *36*, 774–779. [CrossRef] [PubMed]
8. Ohlin, A.; Björkman, L.; Serenius, F.; Schollin, J.; Källén, K. Sepsis as a Risk Factor for Neonatal Morbidity in Extremely Preterm Infants. *Acta Paediatr. Int. J. Paediatr.* **2015**, *104*, 1070–1076. [CrossRef] [PubMed]
9. Cohen-Wolkowiez, M.; Moran, C.; Benjamin, D.K.; Cotten, C.M.; Clark, R.H.; Benjamin, D.K.; Smith, P.B. Early and Late Onset Sepsis in Late Preterm Infants. *Pediatr. Infect. Dis. J.* **2009**, *28*, 1052–1056. [CrossRef]
10. Wang, M.L.; Dorer, D.J.; Fleming, M.P.; Catlin, E.A. Clinical Outcomes of Near-Term Infants. *Pediatrics* **2004**, *114*, 372–376. [CrossRef]
11. Tomashek, K.M.; Shapiro-Mendoza, C.K.; Davidoff, M.J.; Petrini, J.R. Differences in Mortality between Late-Preterm and Term Singleton Infants in the United States, 1995-2002. *J. Pediatr.* **2007**, *151*, 1995–2002. [CrossRef]
12. Hornik, C.P.; Fort, P.; Clark, R.H.; Watt, K.; Benjamin, D.K.; Smith, P.B.; Manzoni, P.; Jacqz-Aigrain, E.; Kaguelidou, F.; Cohen-Wolkowiez, M. Early and Late Onset Sepsis in Very-Low-Birth-Weight Infants from a Large Group of Neonatal Intensive Care Units. *Early Hum. Dev.* **2012**, *88*, S69–S74. [CrossRef]
13. Centers for Disease Control and Prevention (CDC). Perinatal group B streptococcal disease after universal screening recommendations–United States 2003–2005. *MMWR. Morb. Mortal. Wkly. Rep.* **2007**, *56*, 701–705.
14. Centers for Disease Control and Prevention (CDC). Trends in perinatal group B streptococcal disease—United States 2000–2006. *MMWR. Morb. Mortal. Wkly. Rep.* **2009**, *58*, 109–112.
15. Dong, Y.; Speer, C.P. Late-Onset Neonatal Sepsis: Recent Developments. *Arch. Dis. Child. Fetal Neonatal Ed.* **2015**, *100*, F257–F263. [CrossRef]

16. Bizzarro, M.J.; Raskind, C.; Baltimore, R.S.; Gallagher, P.G. Seventy-Five Years of Neonatal Sepsis at Yale: 1928–2003. *Pediatrics* **2005**, *116*, 595–602. [CrossRef]
17. Leal, Y.A.; Álvarez-Nemegyei, J.; Velázquez, J.R.; Rosado-Quiab, U.; Diego-Rodríguez, N.; Paz-Baeza, E.; Dávila-Velázquez, J. Risk Factors and Prognosis for Neonatal Sepsis in Southeastern Mexico: Analysis of a Four-Year Historic Cohort Follow-Up. *BMC Pregnancy Childbirth* **2012**, *12*, 48. [CrossRef]
18. Osrin, D.; Vergnano, S.; Costello, A.; Williams, L. Serious Bacterial Infections in Newborn Infants in Developing Countries. *Curr. Opin. Infect. Dis.* **2004**, *17*, 217–224. [CrossRef]
19. Mukhopadhyay, S.; Puopolo, K.M. Risk Assessment in Neonatal Early Onset Sepsis. *Semin. Perinatol.* **2012**, *36*, 408–415. [CrossRef]
20. Simonsen, K.A.; Anderson-Berry, A.L.; Delair, S.F.; Dele Davies, H. Early-Onset Neonatal Sepsis. *Clin. Microbiol. Rev.* **2014**, *27*, 21–47. [CrossRef] [PubMed]
21. Lu, L.; Li, P.; Pan, T.; Feng, X. Pathogens Responsible for Early-Onset Sepsis in Suzhou, China. *Jpn. J. Infect. Dis.* **2020**, *73*, 148–152. [CrossRef] [PubMed]
22. Itoh, K.; Aihara, H.; Takada, S.; Nishino, M.; Lee, Y.; Negishi, H.; Itoh, H. Clinicopathological Differences between Early-Onset and Late-Onset Sepsis and Pneumonia in Very Low Birth Weight Infants. *Fetal Pediatr. Pathol.* **1990**, *10*, 757–768. [CrossRef] [PubMed]
23. Iroh Tam, P.Y.; Bendel, C.M. Diagnostics for Neonatal Sepsis: Current Approaches and Future Directions. *Pediatr. Res.* **2017**, *82*, 574–583. [CrossRef] [PubMed]
24. Wynn, J.L. Defining Neonatal Sepsis. *Curr. Opin. Pediatr.* **2016**, *28*, 135–140. [CrossRef]
25. Sharma, A.; Thakur, A.; Bhardwaj, C.; Kler, N.; Garg, P.; Singh, M.; Choudhury, S. Potential Biomarkers for Diagnosing Neonatal Sepsis. *Curr. Med. Res. Pract.* **2020**, *10*, 12–17. [CrossRef]
26. Bingol, K. Recent Advances in Targeted and Untargeted Metabolomics by NMR and MS/NMR Methods. *High-Throughput* **2018**, *7*, 9. [CrossRef] [PubMed]
27. Vignoli, A.; Ghini, V.; Meoni, G.; Licari, C.; Takis, P.G.; Tenori, L.; Turano, P.; Luchinat, C. High-Throughput Metabolomics by 1D NMR. *Angew. Chem.-Int. Ed.* **2019**, *58*, 968–994. [CrossRef]
28. Sarafidis, K.; Chatziioannou, A.C.; Thomaidou, A.; Gika, H.; Mikros, E.; Benaki, D.; Diamanti, E.; Agakidis, C.; Raikos, N.; Drossou, V.; et al. Urine Metabolomics in Neonates with Late-Onset Sepsis in a Case-Control Study. *Sci. Rep.* **2017**, *7*, 45506. [CrossRef]
29. Mardegan, V.; Giordano, G.; Stocchero, M.; Pirillo, P.; Poloniato, G.; Donadel, E.; Salvadori, S.; Giaquinto, C.; Priante, E.; Baraldi, E. Untargeted and Targeted Metabolomic Profiling of Preterm Newborns with Earlyonset Sepsis: A Case-Control Study. *Metabolites* **2021**, *11*, 115. [CrossRef]
30. Fanos, V.; Caboni, P.; Corsello, G.; Stronati, M.; Gazzolo, D.; Noto, A.; Lussu, M.; Dessì, A.; Giuffrè, M.; Lacerenza, S.; et al. Urinary 1H-NMR and GC-MS Metabolomics Predicts Early and Late Onset Neonatal Sepsis. *Early Hum. Dev.* **2014**, *90*, 78–83. [CrossRef]
31. Ludwig, C.; Viant, M.R. Two-Dimensional J-Resolved NMR Spectroscopy: Review of a Key Methodology in the Metabolomics Toolbox. *Phytochem. Anal.* **2010**, *21*, 22–32. [CrossRef]
32. Huang, Y.; Yang, Y.; Cai, S.; Chen, Z.; Zhan, H.; Li, C.; Tan, C.; Chen, Z. General Two-Dimensional Absorption-Mode J-Resolved NMR Spectroscopy. *Anal. Chem.* **2017**, *89*, 12646–12651. [CrossRef] [PubMed]
33. Georgakopoulou, I.; Chasapi, S.A.; Bariamis, S.E.; Varvarigou, A.; Spraul, M.; Spyroulias, G.A. Metabolic Changes in Early Neonatal Life: NMR Analysis of the Neonatal Metabolic Profile to Monitor Postnatal Metabolic Adaptations. *Metabolomics* **2020**, *16*, 58. [CrossRef] [PubMed]
34. Pang, Z.; Chong, J.; Zhou, G.; de Lima Morais, D.A.; Chang, L.; Barrette, M.; Gauthier, C.; Jacques, P.É.; Li, S.; Xia, J. MetaboAnalyst 5.0: Narrowing the Gap between Raw Spectra and Functional Insights. *Nucleic Acids Res.* **2021**, *49*, W388–W396. [CrossRef]
35. van den Berg, R.A.; Hoefsloot, H.C.J.; Westerhuis, J.A.; Smilde, A.K.; van der Werf, M.J. Centering, Scaling, and Transformations: Improving the Biological Information Content of Metabolomics Data. *BMC Genom.* **2006**, *7*, 142. [CrossRef]
36. Gaude, E.; Chignola, F.; Spiliotopoulos, D.; Spitaleri, A.; Ghitti, M.; Garcìa-Manteiga, J.M.; Mari, S.; Giovanna, M. muma, an R package for metabolomics univariate and multivariate statistical analysis. *Curr. Metab.* **2013**, *1.2*, 180–189. [CrossRef]
37. Tripathi, N.; Cotten, C.M.; Smith, P.B. Antibiotic Use and Misuse in the Neonatal Intensive Care Unit. *Clin. Perinatol.* **2012**, *39*, 61–68. [CrossRef] [PubMed]
38. Patton, L.; Li, N.; Garrett, T.J.; Ruoss, J.L.; Russell, J.T.; de la Cruz, D.; Bazacliu, C.; Polin, R.A.; Triplett, E.W.; Neu, J. Antibiotics Effects on the Fecal Metabolome in Preterm Infants. *Metabolites* **2020**, *10*, 331. [CrossRef] [PubMed]

Article

Phenotyping Green and Roasted Beans of Nicaraguan Coffea Arabica Varieties Processed with Different Post-Harvest Practices

Gaia Meoni [1,2,3,*], Claudio Luchinat [1,2,3], Enrico Gotti [4], Alejandro Cadena [5] and Leonardo Tenori [1,2,3,*]

1. Magnetic Resonance Center (CERM), University of Florence, 50019 Sesto Fiorentino, Italy; luchinat@cerm.unifi.it
2. Department of Chemistry "Ugo Schiff", University of Florence, 50019 Sesto Fiorentino, Italy
3. Consorzio Interuniversitario Risonanze Magnetiche di Metallo Proteine (CIRMMP), 50019 Sesto Fiorentino, Italy
4. Home Office, Via Alessandro Manzoni 2, 50121 Florence, Italy; egotti.green@gmail.com
5. Caravela Coffee Ltd., 44-48 Waterside Wharf Road n.1, London N1 7UX, UK; alejandro.cadena@caravela.coffee

* Correspondence: meoni@cerm.unifi.it (G.M.); tenori@cerm.unifi.it (L.T.); Tel.: +39-055-4574-281 (G.M. & L.T.)

Abstract: Metabolomic tecniques have already been used to characterize two of the most common coffee species, *C. arabica* and *C. canephora*, but no studies have focused on the characterization of green and roasted coffee varieties of a certain species. We aim to provide, using NMR-based metabolomics, detailed and comprehensive information regarding the compositional differences of seven coffee varieties (*C. arabica*) of green and roasted coffee bean batches from Nicaragua. We also evaluated how different varieties react to the same post-harvest procedures such as fermentation time, type of drying and roasting. The characterization of the metabolomic profile of seven different Arabica varieties (Bourbon-typica), allowed us also to assess the possible use of an NMR spectra of bean aqueous extracts to recognize the farm of origin, even considering different farms from the same geographical area (Nueva Segovia). Here, we also evaluated the effect of post-harvest procedures such as fermentation time and type of drying on green and roasted coffee, suggesting that post-harvest procedures can be responsible for different flavours. This study provides proof of concept for the ability of NMR to phenotype coffee, helping to authenticate and optimise the best way of processing coffee.

Keywords: metabolomics; phenotyping; nuclear magnetic resonance spectroscopy; coffee beans; coffee processing; coffee varieties; post-harvest treatment

Citation: Meoni, G.; Luchinat, C.; Gotti, E.; Cadena, A.; Tenori, L. Phenotyping Green and Roasted Beans of Nicaraguan Coffea Arabica Varieties Processed with Different Post-Harvest Practices. *Appl. Sci.* **2021**, *11*, 11779. https://doi.org/10.3390/app112411779

Academic Editor: Chiara Cavaliere

Received: 28 October 2021
Accepted: 3 December 2021
Published: 11 December 2021

1. Introduction

Green coffee beans are one of the most traded commodities, and coffee is the most consumed beverage after water [1]. Its popularity is due to the attractive organoleptic and energetic characteristics of coffee [2]. The quality of coffee principally derives from the grade of green coffee beans that are influenced by several factors, including genetics, geographic localization, altitude of the plantation, climate, agricultural and postharvest processing factors [3,4]. Moreover, the different processing techniques of coffee beans can impact the final product. Usually, in wet-processed coffee, freshly harvested coffee cherries are de-pulped to remove the skin and most of the fruit around the bean. Then, de-pulped coffee beans are placed in tanks where they can naturally ferment for 12–24 h. This fermentation begins to break down the mucilage, which is a sugary, slimy substance that surrounds the beans. Then, the coffee is dried on courtyards under the direct sun or in shade. The exact implementation of these steps influences the organoleptic properties and the quality of the product [5], which can be described also by the presence and the concentration of certain metabolites (small molecules < 1500 Da) in coffee beans [6]. These differences in metabolites can be therefore used as indicators of coffee quality, and can potentially direct the agronomic and post-harvest procedures to a high-quality grade final product [7].

Metabolomics, the omic science that deals with metabolites the final products of all biochemical reactions occurring in a certain biological system, can be considered an optimal tool to characterize the totality of genetic and environmental interactions and their effect on coffee [8]. Therefore, understanding how different factors affect coffee metabolites could be crucial to improve coffee quality. Most of the commonly used analytical techniques have been extensively applied to characterize the levels of certain specific chemical components (e.g., sugars, caffeine, trigonelline, phenolics, tocopherols, chlorogenic acids and lipids) and the metabolic profiles of the two mostly traded coffee species, *Coffea arabica* and *C. canephora* (Robusta) [7,9–12]. The profiles of Arabica and Robusta species can be distinguished using fatty acids, amino acid enantiomers, caffeine and other xanthine alkaloids, chlorogenic acid concentration, as well as other compounds (furans, phenols, quinic acids, pyridines, biogenic amines, terpenes and steroids) [13]. Arabica coffee accounts for 60% of global production and is preferred by customers due to its distinct flavor and aroma [14]. There are currently more than 50 commercially grown Arabica coffee cultivars but with different traits that can be classified into four groups based on their genetic descriptions: Ethiopian Landrace, Bourbon-Typica group, Introgressed, and F1 Hybrid [15]. However, little is known about the chemical differences between coffee varieties [6,16,17]. Considering that coffee species have demonstrated to react differently to external stimuli, it could be interesting to evaluate how different varieties of the same species and cultivation type, react if exposed to the identical post-harvest conditions or roasting procedure. Mengistu et al. 2020, demonstrated that different coffee varieties of *C. arabica* grown in the north-western highlands of Ethiopia are characterized by different levels of trigonelline, chlorogenic acids and caffeine. However, at present there are no metabolomic based studies determining the fingerprint and the profile of green and roasted beans of different coffee varieties. Here, NMR-based metabolomics is applied to characterize seven different coffee varieties of the same species (*C. arabica*) and the same cultivation type (Bourbon-Typica) localized within the same geographic area of Nicaragua. Moreover, we evaluated how they differently react to the same post-harvest procedure and to the same roasting time and temperature. The experimental design also allowed us to evaluate the differences between the same varieties grown by different farms located within the same territory. The possibility to characterize the profile of coffee of different producers, localized within a restricted geographic area, as previously demonstrated by our group in cow's milk [18], could be of potential interest for precise authentication. This could pave the way for the authentic territorial characterization of specialty coffees. Identifying qualitative attributes and characteristic metabolomic profiles of each producer and promoting transparency concerning its origin could help individual farmers to add value to their products and become more involved in upgrading strategies.

2. Materials and Methods

2.1. Coffee Beans

A total of 36 green coffee beans batches were collected in 2019 from three distinct farms (farm 1: 13°47′11.8″ N 86°32′46.5″ W, 1155 m AMSL (above mean sea level); farm 2: 13°45′08.8″ N 86°29′42.3″ W, 1033 m AMSL; farm 3: 13°44′36.1″ N 86°24′25.0″ W, 696 m AMSL) localized in the Nicaraguan department of Nueva Segovia. Table 1 shows all the characteristics of each batch. A total of seven different coffee varieties (*C. arabica*, Bourbon-Typica) have been collected: catuai rojo (CR, number of coffee batches = 8), maracaturra (MC, n = 4), bourbon (BO, n = 8), caturra (CA, n = 4), pacamara (PA, n = 4), tekesic (TE, n = 4), bourbon rojo (BR, n = 4). Selected varieties are recognized by name according to information provided by the growers. With the aim to evaluate the effect of the different types of green coffee processing, we considered for each variety two different times of fermentation (12 h and 24 h duration), and two drying procedures after full washing, namely "under shade" (Us) and "direct sun" (Ds), see Table 1. Green bean batches (80 g) were roasted by Caravela Ltd. (London, UK) using an IKAWA professional roaster (IKAWA Ltd., London, UK) at 220 °C for 5:30 min.

Table 1. List of the Nicaraguan coffee batches analyzed.

Farm	Variety	Fermentation Time (h)	Drying	Municipality	Batch Code
1	CR	12	Us	Dipilto	346
1	CR	12	Ds	Dipilto	347
1	CR	24	Us	Dipilto	348
1	CR	24	Ds	Dipilto	349
1	MC	12	Us	Dipilto	352
1	MC	12	Ds	Dipilto	353
1	MC	24	Us	Dipilto	350
1	MC	24	Ds	Dipilto	351
1	BO	12	Us	Dipilto	356
1	BO	12	Ds	Dipilto	357
1	BO	24	Us	Dipilto	354
1	BO	24	Ds	Dipilto	355
2	CA	12	Us	Dipilto	337
2	CA	12	Ds	Dipilto	336
2	CA	24	Us	Dipilto	335
2	CA	24	Ds	Dipilto	334
2	PA	12	Us	Dipilto	341
2	PA	12	Ds	Dipilto	340
2	PA	24	Us	Dipilto	339
2	PA	24	Ds	Dipilto	338
2	BO	12	Us	Dipilto	345
2	BO	12	Ds	Dipilto	344
2	BO	24	Us	Dipilto	343
2	BO	24	Ds	Dipilto	342
3	CR	12	Us	Mozonte	362
3	CR	12	Ds	Mozonte	363
3	CR	24	Us	Mozonte	364
3	CR	24	Ds	Mozonte	365
3	TE	12	Us	Mozonte	366
3	TE	12	Ds	Mozonte	367
3	TE	24	Us	Mozonte	368
3	TE	24	Ds	Mozonte	369
3	BR	12	Us	Mozonte	358
3	BR	12	Ds	Mozonte	359
3	BR	24	Us	Mozonte	360
3	BR	24	Ds	Mozonte	361

2.2. NMR Samples

Seven beans for each batch were grounded using a Caso 1830 coffee grinder, which was thoroughly cleaned between the grinding of each sample. A total of ~0.2 g of crushed beans were weighed into 2 mL Eppendorf tubes and 1 mL of ultrapure H_2O (Synergy®, Merck KGaA, Darmstadt, Germany) was added to each sample. Samples were centrifuged 5 min at 14,000 RCF (room temperature) and then incubated at 95 °C in closed 2 mL Eppendorf tubes for 1 h. The aqueous extracts were centrifuged for 5 min at 14,000 RCF at 4 °C to let the solid debris settle. Then, 300 µL of the supernatant were transferred into a new 1.5 mL Eppendorf tube and mixed with 300 µL of phosphate buffer (1.5 M K_2HPO_4, 100% (*v/v*) ^2H_2O, 2 mM NaN_3, 5.8 mM TSP; pH 7.4, all reagents have been purchased by Sigma-Aldrich, Darmstadt, Germany) and vortexed for 20 s. A total of 450 µL of this mixture was transferred in a 4.25 mm NMR tube. Samples were weighted and extracted in five technical replicates for each batch.

2.3. NMR Spectroscopic Analysis and Data ProcessingNMR Data Analysis

One-dimensional (1D) ^{1}H-NMR spectra were measured at 400 MHz using an AVANCE III Bruker spectrometer equipped with a 5 mm BBI 400S1 H-BB-D-05Z probe. The probe temperature was regulated at 300 K and for each spectrum, 64 scans were collected using noesygppsld (Bruker) pulse sequence, a spectral width of 12.47 ppm, a relaxation delay of 4 s and a total acquisition time of 8 min. The receiver gain was set to 203. FIDs were zero-filled and transformed using exponential line broadening (0.6 Hz), resulting in spectra of 16,384 data points. A total of 260 noesygppsld spectra were acquired. Because of the low shimming quality, two NMR spectra of roasted coffee beans (361a and 369b) were removed before the statistical analyses (total of considered spectra: n° green = 180; n° roasted = 178).

2.4. NMR Data Analysis

Resulting NMR spectra were aligned to the TSP signal (0 ppm) and input variables for statistical analyses were generated via variable size binning (green coffee beans spectra divided into 384 buckets, and roasted coffee beans spectra divided into 419 buckets). Each spectrum was segmented into buckets of 0.02 ppm in the range between 0.4 and 10 ppm, except the resonance regions of caffeine (3.2, 3.4, 7.75 ppm) and chlorogenic acid (6.2, 7.0, 7.50 ppm) because of the significant chemical shift changes observable due to their interaction in aqueous solution [19]. Therefore, the buckets of these regions were merged to have the protons of the corresponding molecule into the same bucket window (merged in green buckets: 3.14–3.28, 3.34–3.44, 6.02–6.5, 6.58–7.2, 7.44–7.68, 7.7–7.9; merged roasted buckets: 3.22–3.28, 3.36–3.48, 6.22–6.44, 6.72–7.2, 7.44–7.66, 7.7–7.88; Supplementary Materials Figure S1). Moreover, the region of residual water (4.5 ppm–5.24 ppm) was excluded. Buckets were then normalized to the measured weight of crushed beans, and thereafter, Probabilistic Quotient Normalization (PQN) was applied. The resulting dataset was used to perform multivariate statistical analysis.

A total of 20 metabolites were identified in all the NMR spectra of green coffee, and 29 were identified in all the roasted one. Among them, 15 metabolites are present both in the NMR spectra of green and roasted aqueous extracts. Comprehensively, 34 different metabolites were assigned (Figure 1). Since most of them resonate in crowded regions of the spectrum, where the presence of other signals below certain peaks cannot be excluded, only 15 metabolites in green and 25 metabolites in roasted coffee, corresponding to well defined and resolved peaks in the spectra, were quantitated considering the area under the peaks. Signal identification was performed using a library of NMR spectra of pure organic compounds (AssureNMR 2.2 software, Bruker BioSpin, Karlsruhe, Germany), public databases (e.g., FooDB, n.d.; PhytoHub, n.d., Edmonton, Alberta) storing reference and literature data [7,10,12]. The resulting matrices were used to perform multivariate and univariate data analyses.

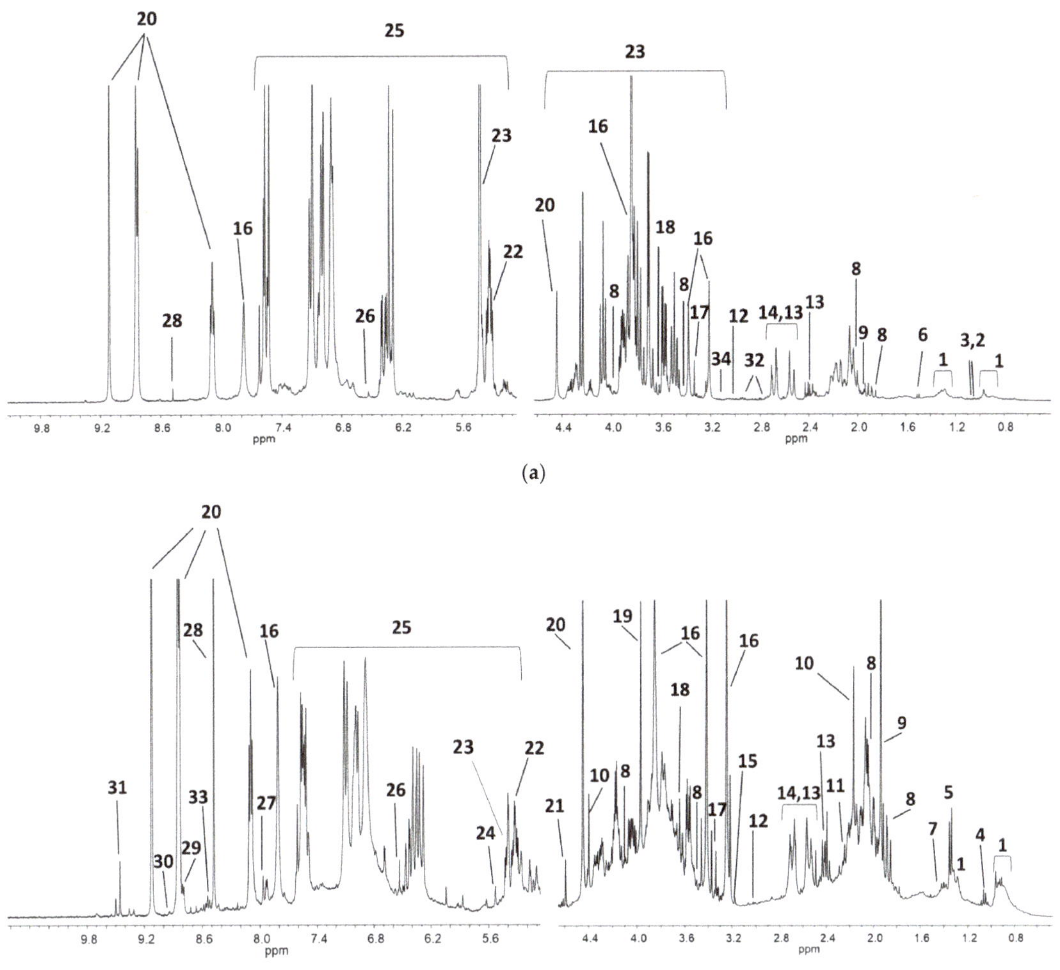

Figure 1. ^{1}H NMR spectra (400 MHz) of water-soluble green (a) and roasted (b) coffee beans. (1) fatty acids; (2) isoleucine; (3) valine; (4) propionic acid; (5) lactic acid; (6) alanine; (7) unknown1; (8) quinic acid; (9) acetic acid; (10) hydroxy acetone; (11) acetone; (12) gamma-aminobutyric acid (GABA); (13) malic acid; (14) citric acid; (15) choline; (16) caffeine; (17) theophylline; (18) myo-inositol; (19) glycolic acid; (20) trigonelline; (21) 2-furylmethanol; (22) 5-caffeolylquinic acid (5-CQA); (23) sucrose; (24) citraconic acid; (25) chlorogenic acids; (26) fumaric acid; (27) xanthine; (28) formic acid; (29) N-methylpyridinium (NMP); (30) nicotinic acid; (31) 5-hydroxymethylfurfural (5-HMF); (32) anserine; (33) methyl xanthines; (34) putrescine.

2.5. Statistical Analysis

Data analyses were performed using R, an open-source software for the statistical analysis of data. Multivariate analysis on metabolomic data was performed on processed NMR bucketed spectra. Principal component analysis (PCA) was used as first exploratory analysis [20]. The RF ("Random Forest" of R package) algorithm [21], was used to assess whether green and roasted NMR metabolomic profiles can be used to classify samples according to the variety, origin, and kind of drying (direct sun or under shadow) and fermentation time (12 h or 24 h) of different coffee batches. Random Forest uses a collection of classification trees, each of them is grown by random selection of features from a bootstrap sample at each branch. Class prediction is based on the majority vote of the collection. While the tree is constructed, about one-third of the instances are left out of the bootstrap sample. This data is then used as test sample to obtain an unbiased estimate of the classification (OOB) error. Variable importance is evaluated by measuring the increase of the OOB error when variables are permuted [22].

Univariate analysis was performed on quantitated metabolites. The Kruskal–Wallis test followed by Dunn post hoc analysis [23] was chosen to infer significant differences among independent samples from multiple groups (n° groups > 2). The Wilcoxon test was chosen to gather differences between two groups and false discovery rate correction was applied using the Benjamini and Hochberg method (FDR) [24], an adjusted p-value < 0.05 was considered statistically significant.

3. Results and Discussion

3.1. Unsupervised Analysis of ^1H NMR Coffee Beans Spectra

As preliminary evaluation, PCA was performed on the datasets of bucketed ^1H-NMR spectra (5 independent samples for each batch), to investigate the quality and the overall behaviour of the acquired green and roasted coffee spectra (Supplementary Materials Figure S2). The sum of the variance of PC1 and PC2 accounts for a total of 89.9% and 76.5% in green and roasted coffee score plots, respectively (Supplementary Materials Figure S2a,b). PCA shows a tendence to form clusters according to the variety (Supplementary Materials Figure S2a,b). The farm effect seems to emerge particularly in BO variety (Supplementary Materials Figure S2a,b), while e subtle differentiation by the fermentation time (12 h vs. 24 h) emerges, especially for the MC and PA green coffee beans water extracts (Supplementary Materials Figure S2a). This is in line with the observation that there are varieties that, being more metabolically susceptible, could also change more significantly in taste depending on the way in which they are processed [24]. Even less marked, these differences are present also in the spectra of roasted beans.

3.2. Coffee Varieties

Each variety was analyzed, using RF as the supervised machine learning approach, to demonstrate the presence of the fingerprint of coffee varieties both in green and in roasted coffee using all NMR data (mean predictive accuracy 91.7%, Table S1). This type of analysis conducted within the same farm, certainly highlights the strong differences between varieties.

Then, the presence of the varietal fingerprint was investigated regardless of the farm of provenance, using bucketed spectra (Figure 2a,b). The predictive accuracies of green (Figure 2a) and roasted (Figure 2b) coffee beans models are similarly good (87.2% in green and 86.0% in roasted), confirming, as previously seen (Supplementary Materials Table S1), that even after roasting the varietal metabolomic fingerprint could be derived. Among all variety classes, TE and BR class error is the highest (Figure 2a,b). The RF variable importance is calculated for green and roasted coffee beans batches, and the overall importance is assessed by determining the maximum for each descriptor over all classes (Figure 2c,d). As shown in Figure 2c,d, there are some conserved regions (ppm), present both in green and roasted coffee spectra, that mostly contribute as important features ranked by RF: the region between 0.94 and 1.2 ppm could be mainly ascribed to the broad signals of methyl and methylene protons of fatty acids (FA) chains [9], the regions within 4.44 and 4.46 ppm, 8.08–8.12 ppm, 8.80–8.86 ppm and 9.12–9.14 ppm, attributable to trigonelline (TR) protons, and the bucket range from 2.52 ppm to 2.76 ppm corresponding to citric acid (CT) signals. Therefore, fatty acids, trigonelline and citric acid, can be considered descriptors of the varieties both in green and in roasted NMR spectra. It emerges that fatty acids and trigonelline maintain also the same trend between the considered varieties in green and in roasted coffee.

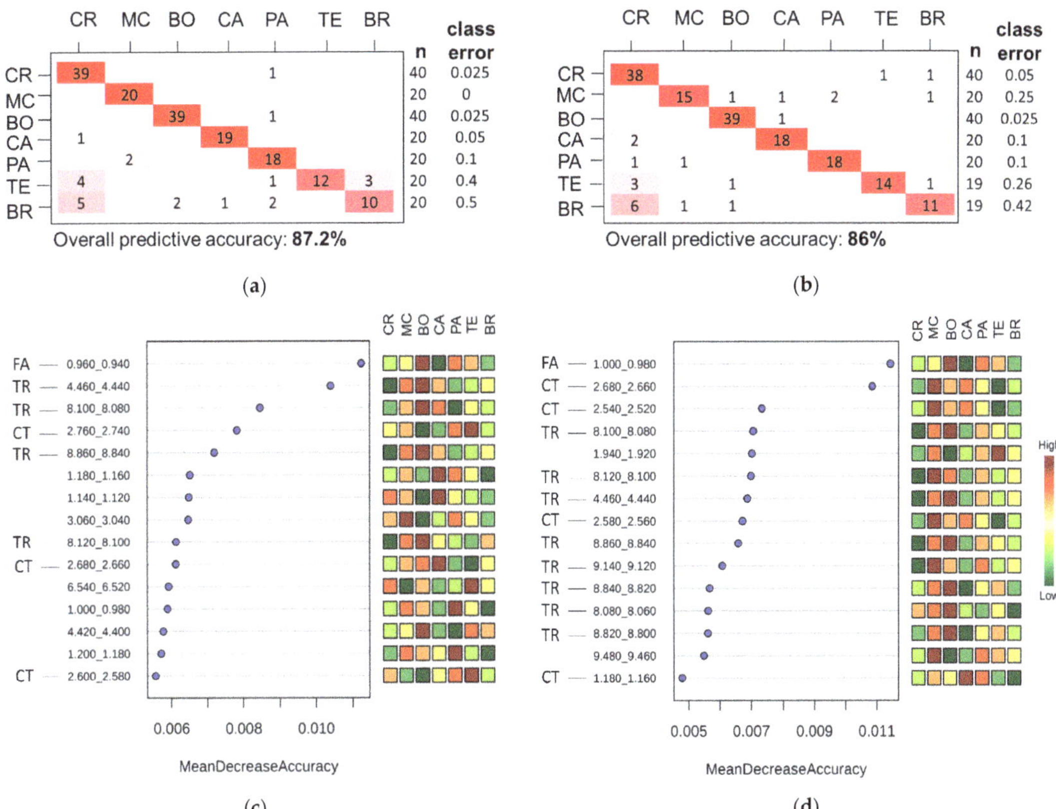

Figure 2. Variety fingerprint assessment through RF. Confusion matrices of RF algorithm of green (**a**) and roasted (**b**) coffee bean spectra. A summary of the variable importance measures for the buckets of coffee NMR spectra with variety as the response variable in the RF model is reported: (**c**) for green coffee model (**a**,**d**) for roasted coffee model (**b**). Buckets are ranked according to the mean decrease in classification accuracy when they are permuted. Calculated RF class error and mean decrease accuracy units can be also read as percentage (e.g., class error of 0.05, means 5%). Most important buckets regions corresponding to assignable resonance present both in green and roasted coffee RF models (**c**,**d**) are labeled accordingly: fatty acid, FA; trigonelline, TR; citric acid, CT. Corresponding RF score plots are reported in Supplementary Materials Figure S3.

In addition, with the aim to compare the efficacy of the fingerprinting and profiling approaches [8], RF was applied, even on the matrices of the corresponding peak areas of the identified metabolites in green and roasted spectra (Figure 3a–d). Compared to the RF models built on bucketed spectra (entire spectra, fingerprinting, Figure 2), models built on metabolites resulted to be less accurate (green model, Figure 3a, pred. acc: 79.4%; roasted model, Figure 3b, pred. acc: 69.7%). This suggests that the fingerprint approach is preferable for variety classification/recognition [8].

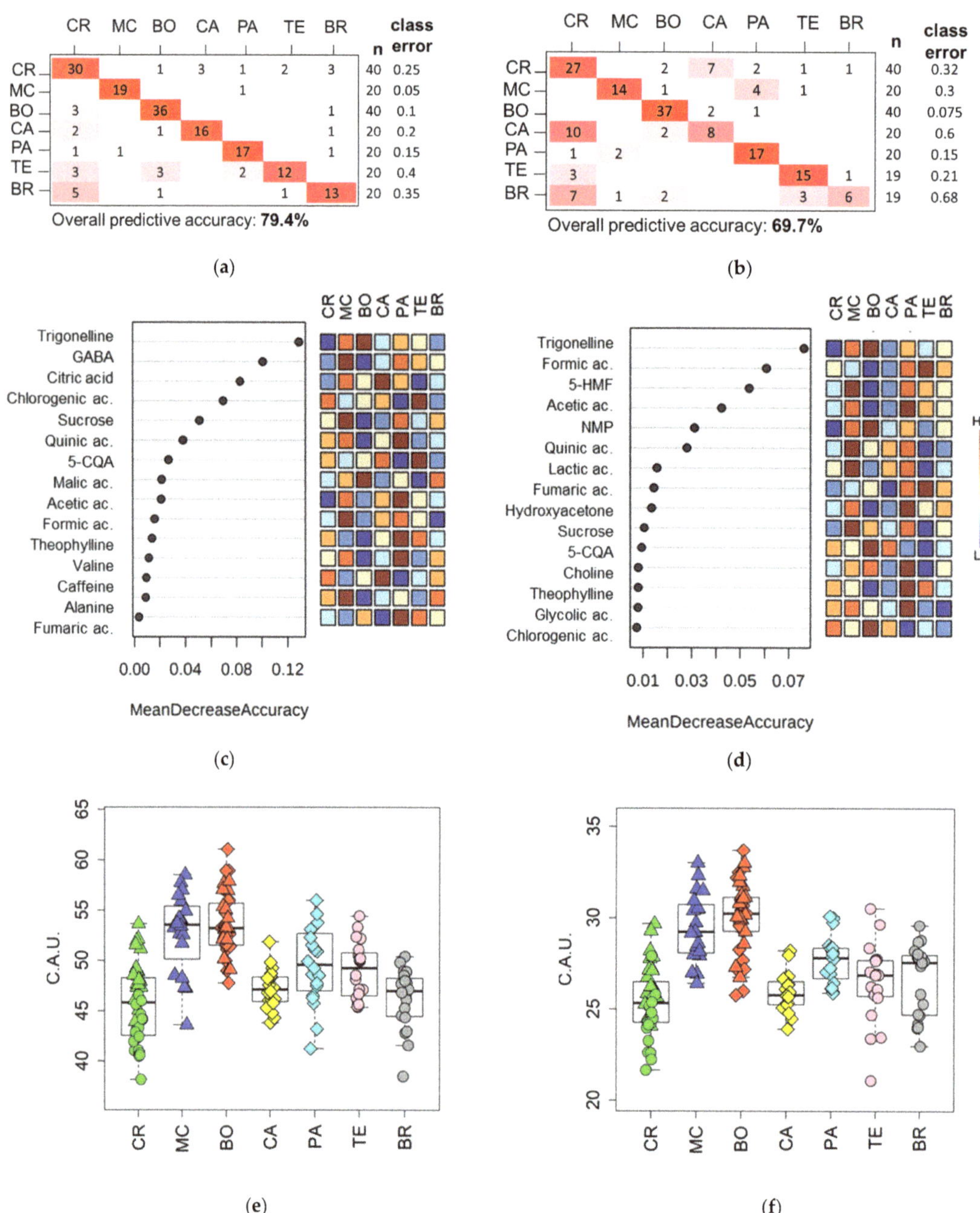

Figure 3. Variety profiling. Confusion matrices of RF algorithm of green (**a**) and roasted (**b**) metabolites of coffee beans. A summary of the variable importance measures for the identified metabolites with variety as the response variable in the RF model (**a**) of green coffee (**c**) and model (**b**) of roasted coffee (**d**). Metabolites are ranked according to the mean decrease in classification accuracy. At the bottom are reported the boxplots of the trigonelline levels in green (**e**) and roasted beans (**f**) among the seven different coffee varieties.

The most important metabolites in discriminating varieties are ranked in Figure 3c,d. This analysis reports trigonelline as the most contributing metabolite in the discrimination of varieties, both in the profile of green (Figure 3c) and in the profile of roasted coffee (Figure 3d).

Univariate statistical analysis (Figure 3e,f, Supplementary Materials Figures S4 and S5) supports that trigonelline trends are conserved between coffee varieties, even after roasting (BO > MC > PA > TE = BR > CA > CR).

An increment of quinic, acetic, fumaric and formic acids and a decrement of gamma-aminobutyric acid (GABA), malic acid, theophylline, trigonelline, 5-CQA, sucrose and chlorogenic acids occur following the roasting process (Supplementary Materials Figure S6). Although it is already known that roasting leads to the alteration of these metabolites [12,25,26], there are no data regarding the behavior of such components among different coffee varieties after roasting. As previously seen, trigonelline is particularly interesting in this respect, as it demonstrated a characteristic trend among varieties and was preservable even after roasting. Characteristically higher amounts of trigonelline are usually detected in *C. arabica* with respect to *C. canephora* [27]. This is in line with reports by Mengistu et al., 2020 on Ethiopian coffee, suggesting a characteristic trend of trigonelline among different varieties. The significance of trigonelline has been well established in previous studies, not only as a precursor of flavor and aroma compounds (as one of the main contributors to coffee's bitter taste), but also as a beneficial nutritional compound [28].

The fact that the trigonelline trend is conserved among varieties after roasting could suggest trigonelline as a potential candidate biomarker for variety determination.

3.3. Coffee Farms

RF models were also created to assess whether the characteristic fingerprint and/or profile of the corresponding coffee farm can be derived from coffee batches of the same variety. However, this hypothesis has been tested only in catuai rojo (CR) and bourbon (BO), since among the seven varieties collected, only these two are produced by more than one farm (see Table 1). RF models have been built to distinguish CR batches of farm 1 and farm 3 and BO batches of farm 1 and farm 2. All the four RF models built to distinguish farms of CR show optimal predictive accuracies (pred. acc%. 94 ± 3.15) both for green and roasted coffee (Supplementary Materials Figure S7a–d). Consistently with the summaries of the most important variables (Supplementary Materials Figure S7a1–d1), univariate analysis on metabolites shows significant higher content of quinic acid, alanine, trigonelline, caffeine, and lower amounts of theophylline, 5-CQA, citric and chlorogenic acid in green coffee beans of the catuai rojo variety of farm 1 when compared to farm 3 (Figure 4a). Higher levels of choline, sucrose, xanthine, and lower levels of 5-hydroxymethyl furfural (5-HMF), fumaric acid, hydroxyacetone and formic acid can be observed in CR roasted coffee of farm 1 (Figure 4b). Even the four RF models built to classify BO coffee batches according to the farm of origin (farm 1 vs. farm 2) show optimal classification accuracies (pred. acc%. 98.1 ± 2.4, Supplementary Materials Figure S7e–h). The summary of variables importance (Supplementary Materials Figure S7f1) and univariate analysis on metabolite levels (Figure 4c) report higher amounts of alanine, 5-CQA, malic and chlorogenic acid, and lower levels of theophylline, quinic acid, GABA and sucrose in green beans of farm 1 when compared to farm 2. The trends of theophylline, 5-CQA and chlorogenic acid are conserved even after roasting (see Figure 4d). Moreover, in roasted BO coffee beans of farm 1 compared to farm 2, lower amounts of 5-HMF, lactic acid, hydroxyacetone, formic and acetic acid and myo-inositol are detected. Taken together, these results suggest that farm 1 is characterized by higher levels of theophylline when compared with the other farms. Theophylline is a xanthine alkaloid and it is usually detected in higher amount in Robusta than in Arabica beans [29,30]. It has already been demonstrated by Mehari et al., that the concentrations of xanthine alkaloids (such as theophylline, theobromine, trigonelline and caffeine) could change significantly in coffee according to geographical origin [31]. Higher levels of theophylline could derive from different caffeine metabolisms of the plant, but

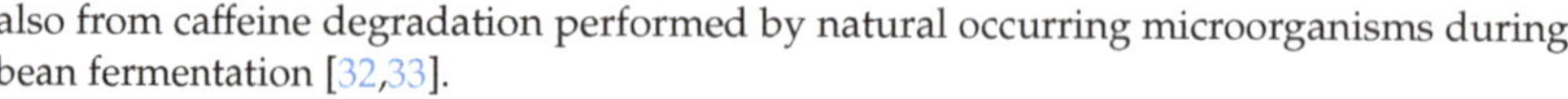

also from caffeine degradation performed by natural occurring microorganisms during bean fermentation [32,33].

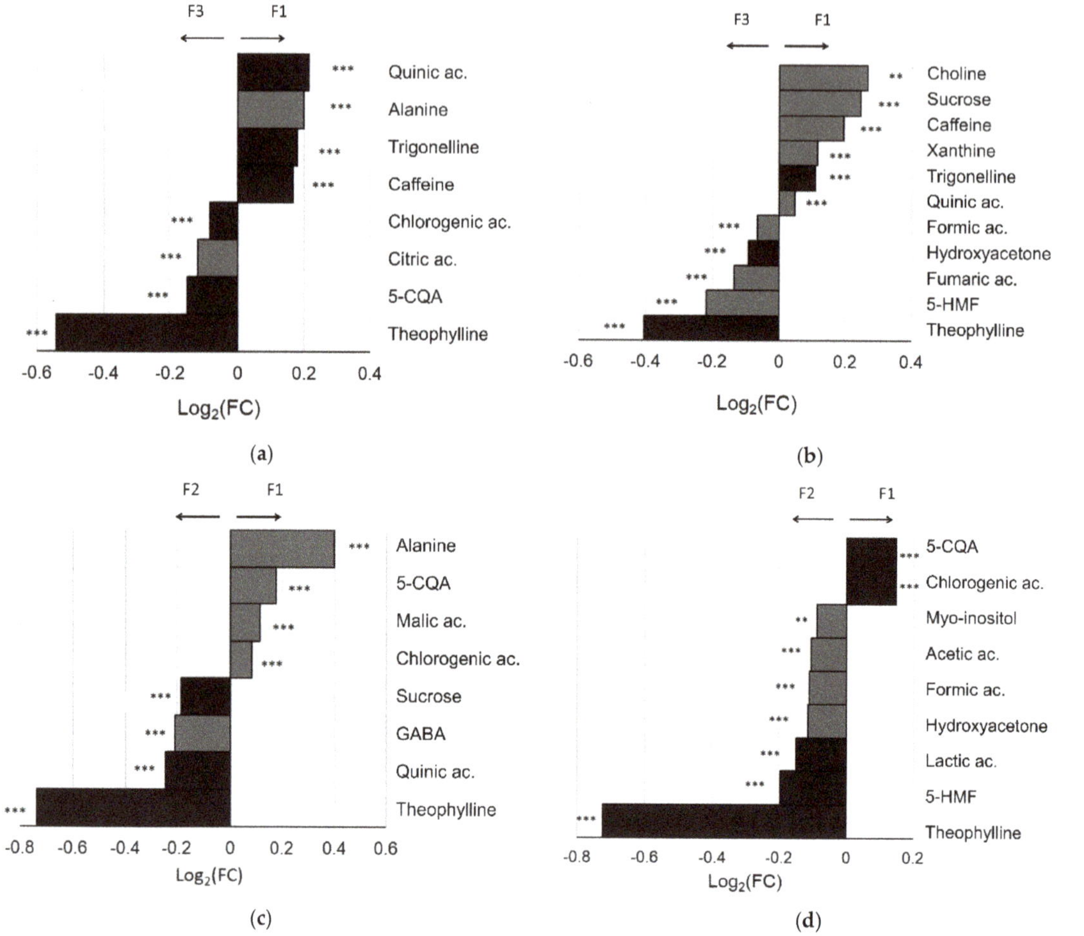

(a) (b) (c) (d)

Figure 4. Log_2(FC) of metabolites' concentrations characteristic of distinct farms. Positive Log_2(FC) values represent higher level of green coffee (**a**) and roasted coffee (**b**) metabolites in farm 1 compared to farm 3 (negative Log_2(FC) values) considering catuai rojo batches; positive values represent higher levels of green coffee (**c**) and roasted coffee (**d**) metabolites in farm 1 compared to farm 2 (negative Log_2(FC) values) considering bourbon batches. Dark gray bars represent the metabolites which are statistically significant after the FDR *p*-value correction (FDR < 0.05), gray bars for those metabolites that show a *p*-value < 0.05 but lose significance after the False Discovery Rate correction. Asterisks represent the Cliff's delta effect-size, were "***" means large effect, "**" medium effect.

3.4. Evaluation of the Fermentation and Drying Effects on Coffee Metabolomic Profile

To evaluate the effect of the two times of fermentation (12 h and 24 h), the RF approach was applied on green and roasted coffees, using either the matrices of bucketed spectra or the matrices of metabolites (Supplementary Materials Figure S8a–d). Considering all four models created, the fermentation time mostly affected the profile of green coffee, while the effect was not remarkable in roasted coffee. The RF model built on green metabolites resulted to be the most effective in discriminating the fermentation times (pred. acc%. 72.2, Supplementary Materials Figure S8b). A total of 24 h fermented coffee beans were characterized by higher levels of acids (in particular: malic acid, acetic acid, chlorogenic acids, 5-CQA, citric acid, fumaric acid, GABA, quinic acid, as reported in Supplementary Materials Figure S8b1).

Univariate analysis corroborates the fact that malic acid levels are statistically different in the two groups. Among all the fifteen quantified metabolites in green coffee beans, only malic acid remained statistically significant after the false discovery rate (FDR) correction (malic acid: p-value = 0.0002, FDR = 0.003, cliff's delta = small). Based on the good performance of the model reported as b in Supplementary Materials Figure S8, to check if distinct varieties reacted differently to 12 h or 24 h of fermentations, the effect was evaluated on green coffee metabolites considering each variety separately. As can be seen in Figure 5, each variety reacts differently to the time of fermentation. In particular, the profiles of fermentation times can be distinguished with a predictive accuracy ~100% in maracaturra, pacamara and bourbon rojo (Figure 5b,e,f), suggesting a remarkable change induced by the time of fermentation. As previously reported, the coffee batches longer fermented are characterized by higher levels of acids.

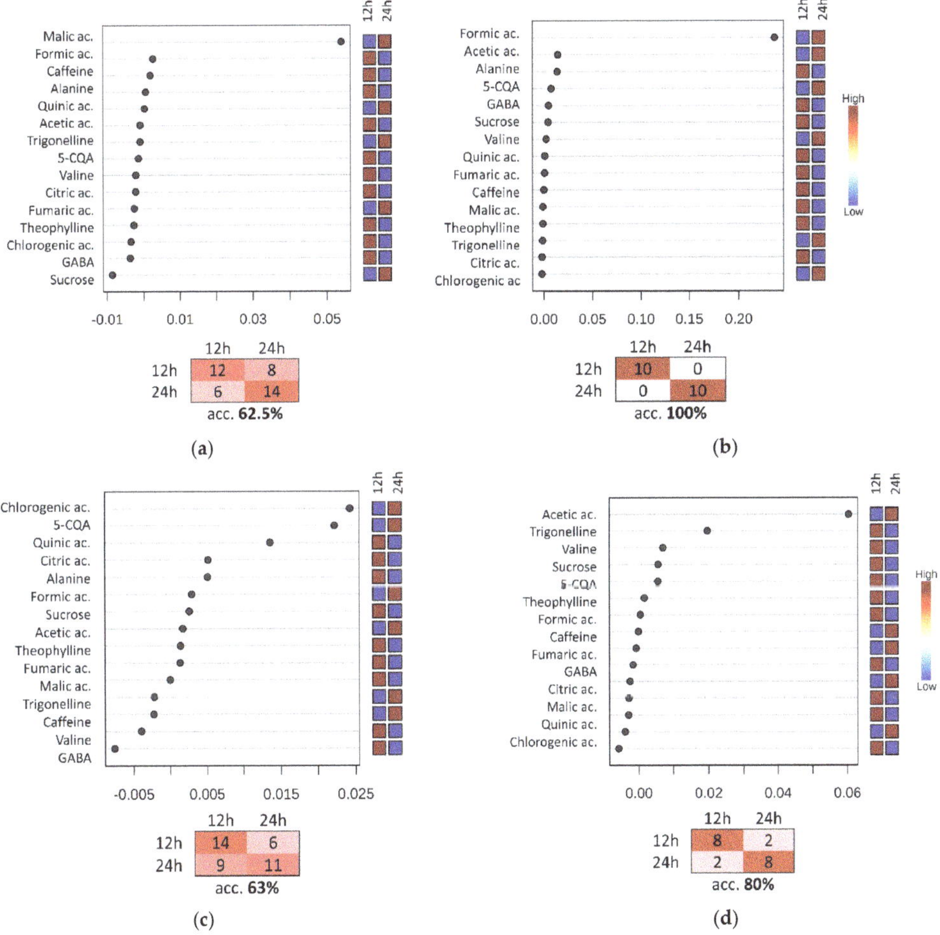

Figure 5. *Cont.*

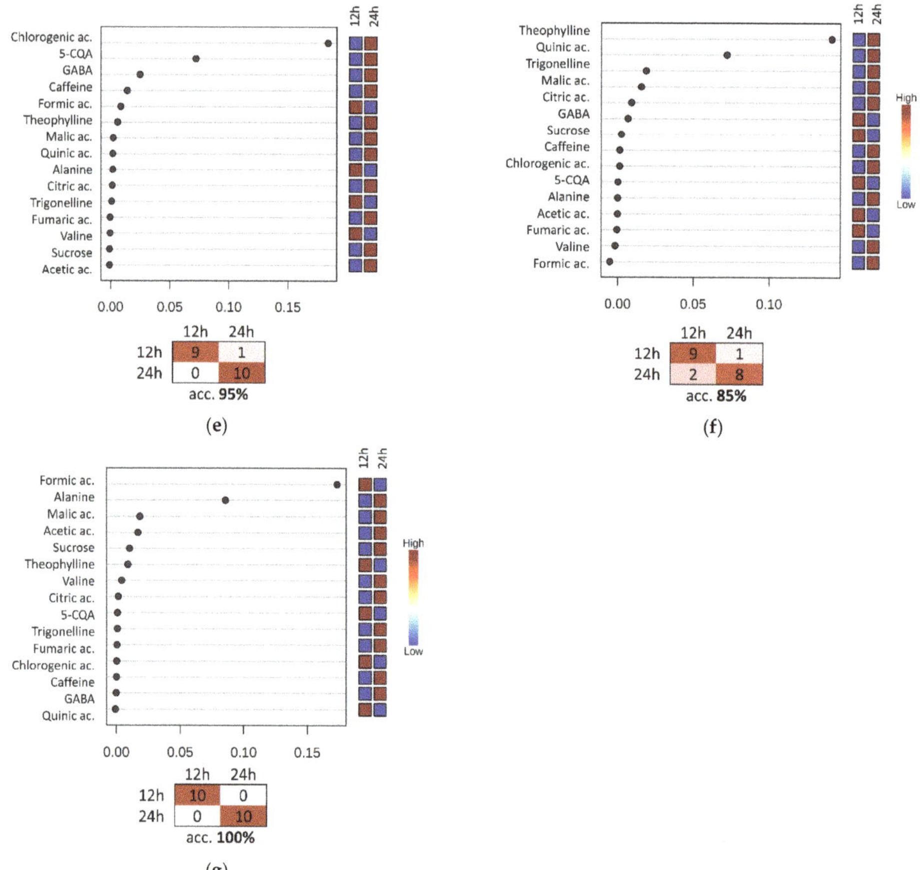

Figure 5. RF models on green coffee metabolites characterizing each variety according to 12 or 24 h of beans fermentation. (**a**) catuai rojo; (**b**) maracaturra; (**c**) bourbon; (**d**) caturra; (**e**) pacamara; (**f**) tekesic; (**g**) bourbon rojo. Mean decrease accuracy values are reported on the orthogonal axis of each plot.

The drying effect was also evaluated considering coffee batches processed "under shade" and those at "direct sun" exposure. Four RF models were created using available data (Supplementary Materials Figure S9a–d): also, in this case the effect of the different drying procedure is more remarkable in green coffee (Supplementary Materials Figure S9a,b), and in particular the RF model built on metabolites provided a better discrimination of green coffee batches processed in the two different manners (overall predictive accuracy: 71%, Supplementary Materials Figure S9b). Among the quantified metabolites, amino acids (valine and alanine) seem to be the most affected by these procedures (Supplementary Materials Figure S9b1). Univariate analysis confirms valine as the only metabolite which remained significantly altered after the FDR correction (valine: p-value = 0.0006, FDR = 0.008, cliff's delta = small). The effect of the drying procedure is detectable in each variety Figure 6. In Figure 6 it emerges that valine levels are higher in all coffee beans batches dried at direct sun. Alanine, acetic and chlorogenic acid levels are generally altered in all the considered models, but there is not a unique trend common for all the varieties (Figure 6a–e), demonstrating that each variety differently reacts to the type of drying.

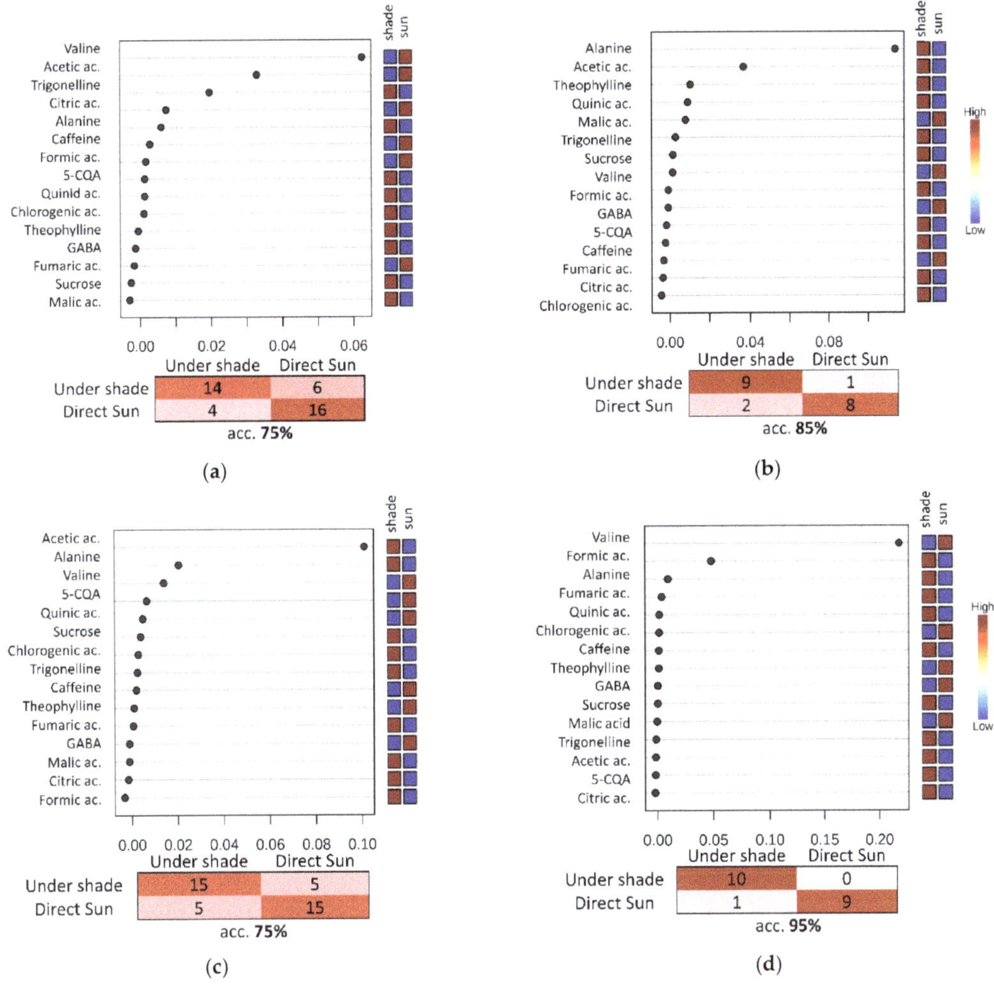

Figure 6. *Cont.*

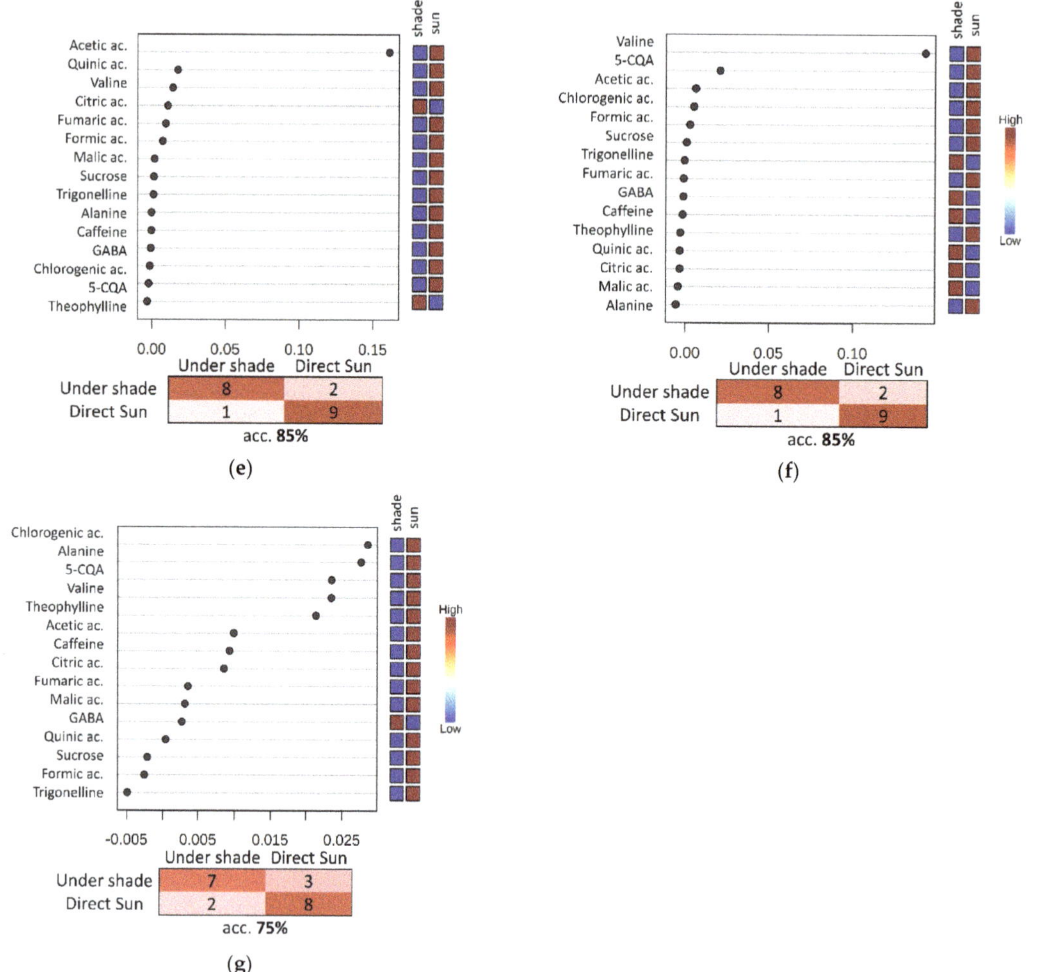

Figure 6. RF models on green coffee metabolites characterizing each variety according to the different drying procedures used: "under shade" and "direct sun". (**a**) catuai rojo; (**b**) maracaturra; (**c**) bourbon; (**d**) caturra; (**e**) pacamara; (**f**) tekesic; (**g**) bourbon rojo. Mean decrease accuracy values are reported on the orthogonal axis of each plot.

4. Conclusions

Coffee metabolomics research has primarily focused on green and roasted coffee beans from the two main varieties, *C. arabica* and *C. canephora.* To the best of our knowledge, there are no metabolomic based studies about the characterization of coffee varieties considering both green and roasted coffee. Here, we have presented detailed and comprehensive information regarding the different metabolomic composition of seven Arabica coffee varieties, using an NMR-based metabolomic approach. For each variety, two points of fermentation time (12 h vs. 24 h) and two types of drying procedures (under shade and direct sun) have been considered. The analyses were performed both considering the entire spectra to evaluate the fingerprint of each variety, and on the identified metabolites, both for green and for roasted coffee beans.

The results demonstrated that NMR spectra of both green and roasted coffee beans can be used to recognise coffee varieties with high accuracies (87.2% and 86% using, respectively green and roasted NMR spectra to build the model).

Moreover, it was also possible to characterize, using this approach, the metabolomic profile of distinct coffee farms within the same restricted geographical area of Nicaragua cultivating the same varieties. Our results demonstrate that, even when coffee batches are processed following the same post-harvest procedure, the characteristic fingerprint of each farm could be derived with high predictive accuracies. The opportunity to quickly obtain NMR spectra with a minimal sample preparation, and to use them to classify samples according to the variety, makes the NMR-based metabolomic approach a suitable approach to recognize original products. Moreover, NMR spectroscopy may be considered as a "magnetic tongue" that analyses and predicts food flavours without being targeted and disruptive.

Therefore, the effects of the time of fermentation and drying types were also evaluated, suggesting that both post-harvest procedures are capable of inducing changes in the metabolic profile of coffee beans that are responsible for different flavours in the cup. In particular, the amount of malic acid, which contributes to a tart acidulous and sour taste, is increased in 24 h of fermentation batches of CR, PA, TE, and BR; trigonelline, instead, related to a bitter taste, is increased in 12 h fermentation in CA, while the other varieties show weaker variation based on the treatments; formic acid which gives a sour/lemon taste, is increased in MC green beans at 24 h of fermentation, while it is decreased in BR cultivar at 24 h of fermentation. Caffeine content seems also to be slightly increased by longer fermentation time. The content of acetic acid, which contributes to a sour vinegar taste, seems to be higher, particularly in CR and PA if exposed to the sun drying, instead, for the other varieties, higher content can be obtained if beans are exposed under shade. The present study suggests that post-harvest treatment procedures can differently affect the amount of aroma precursors within distinct coffee varieties and that the kind of processing should be optimized specifically for each variety.

Supplementary Materials: The following are available online at https://www.mdpi.com/article/10.3390/app112411779/s1, Figure S1: ^{1}H NMR spectra of coffee beans. Figure S2: First two components of PCA score plots of ^{1}H NMR bucketed spectra of green (a) and roasted (b) coffee beans. Figure S3: Random Forest multidimensional scaling (MDS) plots. Figure S4: Univariate analysis on metabolites: cultivars' comparison (green coffee beans). Figure S5: Univariate analysis on metabolites: cultivars' comparison (roasted coffee beans). Figure S6: Roasting effect on coffee beans' metabolites. Figure S7: Farms' fingerprint and profiling assessment through RF. Figure S8: RF models built on green and roasted NMR bucketed spectra and metabolites: beans fermented 12 h vs. beans fermented 24 h. Figure S9: RF models built on green and roasted NMR bucketed spectra and metabolites: beans dried under shade vs. beans dried at direct sun.Supplementary data tables: containing NMR data (bucketed spectra and area under the peaks of identified molecules) of green and roasted coffee batches used in this study. Table S1: Variety classification within the three farms through RF.

Author Contributions: Conceptualization G.M., C.L., E.G. and L.T.; Data curation G.M.; Formal analysis G.M.; Investigation G.M. and L.T.; Visualization G.M.; Writing Original Draft G.M.; Supervision L.T. and C.L.; Provision of samples A.C.; Funding acquisition C.L.; writing review L.T. and C.L. All authors have read and agreed to the published version of the manuscript.

Funding: This research received no external funding.

Institutional Review Board Statement: Not applicable.

Informed Consent Statement: Not applicable.

Data Availability Statement: Data are contained within the supplementary data table.

Conflicts of Interest: The authors declare no conflict of interest.

References

1. Vegro, C.L.R.; de Almeida, L.F. Chapter 1—Global Coffee Market: Socio-Economic and Cultural Dynamics. In *Coffee Consumption and Industry Strategies in Brazil*; De Almeida, L.F., Spers, E.E., Eds.; Woodhead Publishing Series in Consumer Sci & Strat Market; Woodhead Publishing: Sawston, UK, 2020; pp. 3–19. ISBN 978-0-12-814721-4.
2. Samoggia, A.; Riedel, B. Consumers' Perceptions of Coffee Health Benefits and Motives for Coffee Consumption and Purchasing. *Nutrients* **2019**, *11*, 653. [CrossRef]
3. Hameed, A.; Hussain, S.A.; Ijaz, M.U.; Ullah, S.; Pasha, I.; Suleria, H.A.R. Farm to Consumer: Factors Affecting the Organoleptic Characteristics of Coffee. II: Postharvest Processing Factors. *Compr. Rev. Food Sci. Food Saf.* **2018**, *17*, 1184–1237. [CrossRef]
4. Worku, M.; de Meulenaer, B.; Duchateau, L.; Boeckx, P. Effect of Altitude on Biochemical Composition and Quality of Green Arabica Coffee Beans Can Be Affected by Shade and Postharvest Processing Method. *Food Res. Int.* **2018**, *105*, 278–285. [CrossRef]
5. De Bruyn, F.; Zhang, S.J.; Pothakos, V.; Torres, J.; Lambot, C.; Moroni, A.V.; Callanan, M.; Sybesma, W.; Weckx, S.; De Vuyst, L. Exploring the Impacts of Postharvest Processing on the Microbiota and Metabolite Profiles during Green Coffee Bean Production. *Appl. Environ. Microbiol.* **2017**, *83*, e02398-16. [CrossRef]
6. Guizellini, F.C.; Marcheafave, G.G.; Rakocevic, M.; Bruns, R.E.; Scarminio, I.S.; Soares, P.K. PARAFAC HPLC-DAD Metabolomic Fingerprint Investigation of Reference and Crossed Coffees. *Food Res. Int.* **2018**, *113*, 9–17. [CrossRef]
7. Kwon, D.-J.; Jeong, H.-J.; Moon, H.; Kim, H.-N.; Cho, J.-H.; Lee, J.-E.; Hong, K.S.; Hong, Y.-S. Assessment of Green Coffee Bean Metabolites Dependent on Coffee Quality Using a 1H NMR-Based Metabolomics Approach. *Food Res. Int.* **2015**, *67*, 175–182. [CrossRef]
8. Vignoli, A.; Ghini, V.; Meoni, G.; Licari, C.; Takis, P.G.; Tenori, L.; Turano, P.; Luchinat, C. High-Throughput Metabolomics by 1D NMR. *Angew. Chem. Int. Ed. Engl.* **2019**, *58*, 968–994. [CrossRef]
9. Consonni, R.; Cagliani, L.R.; Cogliati, C. NMR Based Geographical Characterization of Roasted Coffee. *Talanta* **2012**, *88*, 420–426. [CrossRef]
10. Wei, F.; Furihata, K.; Hu, F.; Miyakawa, T.; Tanokura, M. Complex Mixture Analysis of Organic Compounds in Green Coffee Bean Extract by Two-Dimensional NMR Spectroscopy. *Magn. Reson. Chem.* **2010**, *48*, 857–865. [CrossRef]
11. Wei, F.; Furihata, K.; Koda, M.; Hu, F.; Kato, R.; Miyakawa, T.; Tanokura, M. 13C NMR-Based Metabolomics for the Classification of Green Coffee Beans According to Variety and Origin. *J. Agric. Food Chem.* **2012**, *60*, 10118–10125. [CrossRef]
12. Wei, F.; Furihata, K.; Koda, M.; Hu, F.; Miyakawa, T.; Tanokura, M. Roasting Process of Coffee Beans as Studied by Nuclear Magnetic Resonance: Time Course of Changes in Composition. *J. Agric. Food Chem.* **2012**, *60*, 1005–1012. [CrossRef]
13. Monakhova, Y.B.; Ruge, W.; Kuballa, T.; Ilse, M.; Winkelmann, O.; Diehl, B.; Thomas, F.; Lachenmeier, D.W. Rapid Approach to Identify the Presence of Arabica and Robusta Species in Coffee Using 1H NMR Spectroscopy. *Food Chem.* **2015**, *182*, 178–184. [CrossRef]
14. Bicho, N.C.; Leitão, A.E.; Ramalho, J.C.; de Alvarenga, N.B.; Lidon, F.C. Impact of Roasting Time on the Sensory Profile of Arabica and Robusta Coffee. *Ecol. Food Nutr.* **2013**, *52*, 163–177. [CrossRef]
15. World Coffee Research. Available online: https://worldcoffeeresearch.org/ (accessed on 26 November 2021).
16. Mengistu, M.W.; Workie, M.A.; Mohammed, A.S. Biochemical Compounds of Arabica Coffee (*Coffea arabica* L.) Varieties Grown in Northwestern Highlands of Ethiopia. *Cogent Food Agric.* **2020**, *6*, 1741319. [CrossRef]
17. Wang, Y.; Wang, X.; Hu, G.; Hong, D.; Bai, X.; Guo, T.; Zhou, H.; Li, J.; Qiu, M. Chemical Ingredients Characterization Basing on 1H NMR and SHS-GC/MS in Twelve Cultivars of Coffea Arabica Roasted Beans. *Food Res. Int.* **2021**, *147*, 110544. [CrossRef]
18. Tenori, L.; Santucci, C.; Meoni, G.; Morrocchi, V.; Matteucci, G.; Luchinat, C. NMR Metabolomic Fingerprinting Distinguishes Milk from Different Farms. *Food Res. Int.* **2018**, *113*, 131–139. [CrossRef]
19. D'Amelio, N.; Fontanive, L.; Uggeri, F.; Suggi-Liverani, F.; Navarini, L. NMR Reinvestigation of the Caffeine–Chlorogenate Complex in Aqueous Solution and in Coffee Brews. *Food Biophys.* **2009**, *4*, 321–330. [CrossRef]
20. Abdi, H.; Williams, L.J. Principal Component Analysis. *WIREs Comp. Stat.* **2010**, *2*, 433–459. [CrossRef]
21. Breiman, L. Random Forests. *Mach. Learn.* **2001**, *45*, 5–32. [CrossRef]
22. Liaw, A.; Wiener, M. Classification and Regression by RandomForest. *R News* **2002**, *2*, 18–22.
23. Posthoc.Kruskal.Dunn.Test: Pairwise Test for Multiple Comparisons of Mean Rank Sums (Dunn's-Test). Available online: https://www.rdocumentation.org/packages/PMCMR/versions/4.3/topics/posthoc.kruskal.dunn.test (accessed on 27 November 2021).
24. Ribeiro, B.B.; de Carvalho, A.M.; Cirillo, M.Â.; Câmara, F.M.d.M.; Montanari, F.F. Sensory Profile of Coffees of Different Cultivars, Plant Exposure and Post-Harvest. *Afr. J. Agric. Res.* **2019**, *14*, 1111–1113. [CrossRef]
25. Casal, S.; Oliveira, M.B.P.P.; Alves, M.R.; Ferreira, M.A. Discriminate Analysis of Roasted Coffee Varieties for Trigonelline, Nicotinic Acid, and Caffeine Content. *J. Agric. Food Chem.* **2000**, *48*, 3420–3424. [CrossRef]
26. Farag, M.A.; El-Kersh, D.M.; Ehrlich, A.; Choucry, M.A.; El-Seedi, H.; Frolov, A.; Wessjohann, L.A. Variation in Ceratonia Siliqua Pod Metabolome in Context of Its Different Geographical Origin, Ripening Stage and Roasting Process. *Food Chem.* **2019**, *283*, 675–687. [CrossRef]
27. Campa, C.; Ballester, J.F.; Doulbeau, S.; Dussert, S.; Hamon, S.; Noirot, M. Trigonelline and Sucrose Diversity in Wild Coffea Species. *Food Chem.* **2004**, *88*, 39–43. [CrossRef]
28. Wei, F.; Tanokura, M. Chapter 10—Chemical Changes in the Components of Coffee Beans during Roasting. In *Coffee in Health and Disease Prevention*; Preedy, V.R., Ed.; Academic Press: San Diego, CA, USA, 2015; pp. 83–91. ISBN 978-0-12-409517-5.

29. Alonso-Salces, R.M.; Serra, F.; Reniero, F.; HÉberger, K. Botanical and Geographical Characterization of Green Coffee (Coffea Arabica and Coffea Canephora): Chemometric Evaluation of Phenolic and Methylxanthine Contents. *J. Agric. Food Chem.* **2009**, *57*, 4224–4235. [CrossRef]
30. Clifford, M.N.; Kazi, T. The Influence of Coffee Bean Maturity on the Content of Chlorogenic Acids, Caffeine and Trigonelline. *Food Chem.* **1987**, *26*, 59–69. [CrossRef]
31. Mehari, B.; Redi-Abshiro, M.; Chandravanshi, B.S.; Atlabachew, M.; Combrinck, S.; McCrindle, R. Simultaneous Determination of Alkaloids in Green Coffee Beans from Ethiopia: Chemometric Evaluation of Geographical Origin. *Food Anal. Methods* **2016**, *9*, 1627–1637. [CrossRef]
32. Jeszka-Skowron, M.; Frankowski, R.; Zgoła-Grześkowiak, A. Comparison of Methylxantines, Trigonelline, Nicotinic Acid and Nicotinamide Contents in Brews of Green and Processed Arabica and Robusta Coffee Beans—Influence of Steaming, Decaffeination and Roasting Processes on Coffee Beans. *LWT* **2020**, *125*, 109344. [CrossRef]
33. Zhou, B.; Ma, C.; Ren, X.; Xia, T.; Li, X. LC–MS/MS-Based Metabolomic Analysis of Caffeine-Degrading Fungus Aspergillus Sydowii during Tea Fermentation. *J. Food Sci.* **2020**, *85*, 477–485. [CrossRef]

applied sciences

Review

NMR in Metabolomics: From Conventional Statistics to Machine Learning and Neural Network Approaches

Carmelo Corsaro [1,*], Sebastiano Vasi [1], Fortunato Neri [1], Angela Maria Mezzasalma [1], Giulia Neri [2] and Enza Fazio [1]

[1] Department of Mathematical and Computational Science, Physical Science and Earth Science, University of Messina, Viale F. Stagno D'Alcontres 31, I-98166 Messina, Italy; vasis@unime.it (S.V.); fneri@unime.it (F.N.); angelamaria.mezzasalma@unime.it (A.M.M.); enfazio@unime.it (E.F.)

[2] Department of Chemical, Biological, Pharmaceutical and Environmental Sciences, University of Messina, Viale F. Stagno D'Alcontres 31, I-98166 Messina, Italy; giulia.neri@unime.it

* Correspondence: ccorsaro@unime.it

Abstract: NMR measurements combined with chemometrics allow achieving a great amount of information for the identification of potential biomarkers responsible for a precise metabolic pathway. These kinds of data are useful in different fields, ranging from food to biomedical fields, including health science. The investigation of the whole set of metabolites in a sample, representing its fingerprint in the considered condition, is known as metabolomics and may take advantage of different statistical tools. The new frontier is to adopt self-learning techniques to enhance clustering or classification actions that can improve the predictive power over large amounts of data. Although machine learning is already employed in metabolomics, deep learning and artificial neural networks approaches were only recently successfully applied. In this work, we give an overview of the statistical approaches underlying the wide range of opportunities that machine learning and neural networks allow to perform with accurate metabolites assignment and quantification.Various actual challenges are discussed, such as proper metabolomics, deep learning architectures and model accuracy.

Keywords: NMR; metabolomics; biomarkers; clustering; artificial intelligence; machine learning; deep learning; health science

Citation: Corsaro, C.; Vasi, S.; Neri, F.; Mezzasalma, A.M.; Neri, G.; Fazio, E. NMR in Metabolomics: From Conventional Statistics to Machine Learning and Neural Network Approaches. *Appl. Sci.* **2022**, *12*, 2824. https://doi.org/10.3390/app12062824

Academic Editor: Alessia Vignoli

Received: 31 January 2022
Accepted: 1 March 2022
Published: 9 March 2022

Publisher's Note: MDPI stays neutral with regard to jurisdictional claims in published maps and institutional affiliations.

1. Introduction

Metabolomics corresponds to the part of omics sciences that investigates the whole set of small molecule metabolites in an organism, representing a large number of compounds, such as a portion of organic acids, amino acids, carbohydrates, lipids, etc. [1–3]. The investigation and the recording of metabolites by target analysis, metabolic profiling and metabolic fingerprinting (i.e., extracellular metabolites) are fundamental steps for the discovery of biomarkers, helping in diagnoses and designing appropriate approaches for drug treatment of diseases [4,5]. There are many databases available with metabolomics data, including spectra acquired by nuclear magnetic resonance (NMR) and mass spectrometry (MS), but also metabolic pathways. Among them, we mention the Human Metabolome Database (HMDB) [6] and Biological Magnetic Resonance Bank (BMRB) [7] that contain information on a large number of metabolites gathered from different sources. By means of the corresponding web platform, it is possible, for instance, to search for mono- and bi-dimensional spectra of metabolites, starting from their peak position [3]. However, metabolomics databases still lack homogeneity mainly due to the different acquisition conditions, including employed instruments. Thus, the definition of uniform and minimum reporting standards and data formats would allow an easier comparison and a more accurate investigation of metabolomics data [8].

In recent years, NMR has become one of the most employed analytical non-destructive techniques for clinical metabolomics studies. In fact, it allows to detect and quantify metabolic components of a biological matrix whose concentration is comparable or bigger than 1 μM (see Appendix A). Such sensitivity, relatively low if compared with other MS techniques, allows to assign up to 20 metabolites in vivo, and up to 100 metabolites in vitro [9–11]. Numerous strategies are being designed to overcome actual limitations, including a lower selectivity compared to the MS technique coupled with gas or liquid chromatography (GC-MS and LC-MS, respectively) and a low resolution for complex biological matrices. These include the development of new pulse sequences mainly involving field gradients for observing multidimensional hetero- or homo-nuclear correlations [12]. Within metabolomics investigations, NMR analyses are usually coupled with statistical approaches: sample randomization allows to reduce the correlation between confounding variables, sample investigation order and experimental procedures. In the last ten years, nested stratified proportional randomization and matched case-control design were adopted in the case of imbalanced results [13–15].

In any case, data pre-processing is a relevant step before performing data analysis by means of a conventional approach or a statistical one. The goal of pre-processing is to homogenize the acquired data, avoiding the presence of instrumental bias mainly involving peaks' features for a better quantification of metabolites. For example, the pre-processing of NMR spectra concerns phasing, baseline correction, peak alignment, apodization procedures, normalization and binning [16,17] (see Figure 1). In particular, the binning procedure corresponding to the spectral segmentation is performed mainly in those cases of challenging spectral alignment or simply for reducing the data points [18]. Even though binning reduces data resolution, binning procedures are commonly used [19–21].

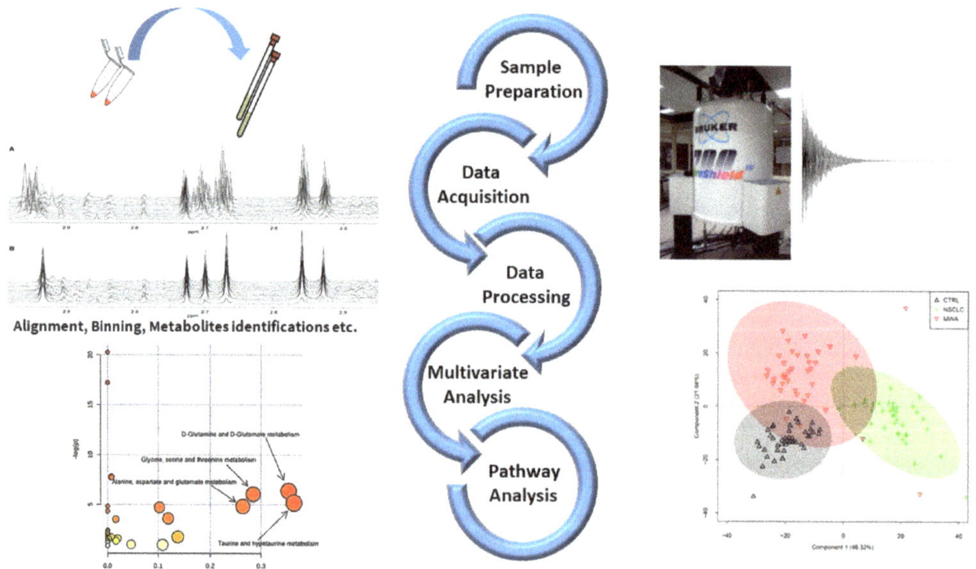

Figure 1. Schematic workflow illustrating the steps of NMR based metabolomic studies coupled with chemometrics and pathway analysis. (1) Sample preparation and NMR tube filling (top left); (2) experimental parameters setting and data acquisition (top right); (3) data processing (middle left); (4) execution of multivariate statistical analysis (bottom right); (5) determination of metabolic pathways (bottom left). Some figures are reprinted from Refs. [22,23] under the terms of the CC-BY license.

For what concerns normalization, recorded spectra are usually normalized by the total integrated area and thus the metabolites concentration can be compared among different samples. In the case of large signals variation, probabilistic quotient normalization can be adopted instead [24]. Finally, deconvolution is also employed for the necessary assignment and quantification of those metabolites whose signals overlap [25,26]. All these pre-processing methods are also chosen, taking into account that the approaches adopted for the data processing are essentially dual: (1) chemometrics, consisting in the employment of statistical analysis for the recognition of similar patterns and for the significant determination of intensity values, and (2) quantitative metabolomics, based on an initial assignment and quantification of metabolites with the subsequent statistics. We outline that, from one side, chemometrics allows an automatic and non-biased classification of metabolites, whereas from the other side, it needs a big number of uniform spectra. These requirements do not apply for quantitative metabolomics [27,28].

In order to gain useful insights and a corresponding interpretation of NMR outcomes, it is indeed mandatory to use statistical and bioinformatic tools, considering the complex output generated [22]. In this work, we discuss the main statistical approaches currently used for NMR-based metabolomics analysis, pointing out the advantages and disadvantages. Illustrative examples are reported, and the actual challenges influencing the analysis are also discussed. On the basis of these evidences, it emerged that innovative experimental procedures would need to be implemented in order to improve the potentiality of existing approaches (i.e., adequate sample sizes and conditions), thereby combining their complementing features with the aim to achieve most of the metabolomic information from an NMR measurement. Nevertheless, on considering the high complexity of biological systems, each regulation level, including the genome, should be considered, yielding corresponding insights on cellular processes. Thus, data coming from different biological levels should be integrated within the same analysis for the observation of interconnectivity changes between the different cellular components. In this context, neural network-based approaches could be adequate in responding to this major challenge and indeed to the exploitation of proper approaches for the weighted consideration of data corresponding to different layers of biological organization.

2. Conventional Approaches

2.1. Unsupervised Methods

In the analysis of large metabolomic NMR datasets, unsupervised techniques are applied with the aim to identify any significant pattern within unlabeled databases without any human action. Below, we introduce and describe several unsupervised methods, highlighting their characteristics and implementation procedures. In particular, we describe the following unsupervised techniques: (a) principal component analysis (PCA); (b) clustering; (c) self-organizing maps (SOMs).

2.1.1. Principal Component Analysis (PCA)

Principal component analysis (PCA) is employed for lowering the dimensionality of high-dimensional datasets, preserving as much information as possible by means of a "linear" multivariate analysis [29,30]. This approach employs a linear transformation to define a new smaller set of "summary indices"—or "principal components" (PCs)—that are more easily visualized and analyzed [31]. In this frame, principal components correspond to new variables obtained by the linear combination of the initial variables by solving an eigenvalue/eigenvector problem. The first principal component (PC1) represents the "path" along which the variance of the data is maximized. As happens for the first principal component, the second one (PC2) also defines the maximum variance in the database. Nevertheless, it is completely uncorrelated to the PC1 following a direction that is orthogonal to the first component path. This step reiterates based on the dimensionality of the system, where a next principal component is the direction orthogonal to the prior components with the most variance. If there are significant distinctions between the ranges of initial variables (those variables with smaller ranges will be dominated by those with larger ones), distorted results may occur. To avoid this kind of problem, it is required to perform a standardization operation before executing PCA that corresponds to a transformation of the data into comparable scales. This can be done by using different scaling transformations, such as autoscaling, the generalized logarithm transform or the Pareto scaling with the aim to enhance the importance of small NMR signals, whose variation is more affected by the noise [32]. One of the most used transformation is the mean centered autoscaling:

$$\frac{value - mean}{st.deviation}, \tag{1}$$

Furthermore, the computation of the covariance matrix is required to discard redundant information mainly due to the presence of any relationship between the initial variables of the data. The covariance matrix is symmetric (a $n \times n$) being composed by the covariances of all pairs of the considered n variables $(x_1, ..., x_n)$:

$$\begin{bmatrix} Cov(x_1, x_1) & \cdots & Cov(x_1, x_n) \\ \vdots & \ddots & \vdots \\ Cov(x_n, x_1) & \cdots & Cov(x_n, x_n) \end{bmatrix} \tag{2}$$

In this frame, PCs can be obtained by finding the eigenvectors and eigenvalues from this covariance matrix. Figure 2 shows a graph with only three variables axes of the n-dimensional variables space. The red point in this figure represents the average point used to move the origin of the coordinate system by means of the mean-centering procedure in the standardization process. Once we define PC1 and PC2, as shown in Figure 2, they define a plane that allows inspecting the organization of the studied database. Further, the projection of the data with respect to the new variables (PCs) is called the score plot, and if the data are statistically different/similar, they can be regrouped and classified.

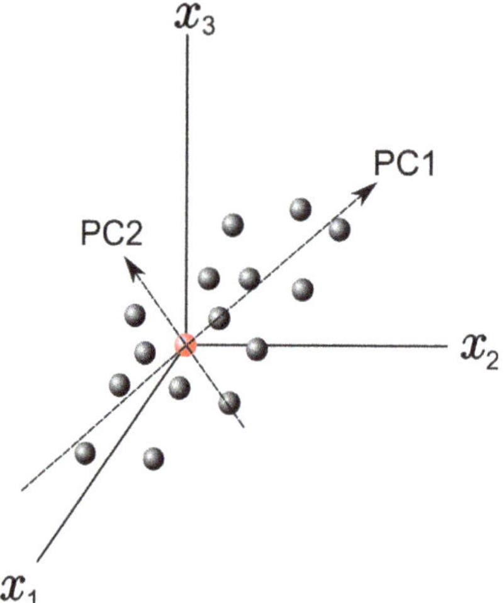

Figure 2. Example plot with 3 variable axes in a n-dimensional variable space. The principal components PC1 and PC2 are reported.

PCA is usually applied in NMR-metabolomic studies because it simplifies the investigation of hundreds of thousands of chemical components in metabolomic database composed of several collected NMR spectra. In this way, each NMR spectrum is confined to a single point in the score plot in which similar spectra are regrouped, and differences on the PC axes shed light on experimental variations between the measurements [28,33–35].

However, it is noteworthy that PCA, like the other latent structure techniques, must be applied to matrices where the number of cases is greater than the number of variables [36].

The PCA technique can also be combined with other statistical approaches, including the analysis of variance (ANOVA) as reported by Smilde et al. [37] in their ANOVA-simultaneous component analysis (ASCA). This method is able to associate observed data changes to the different experimental designs. It is applied to metabolomics data, for example, to study variations of the metabolites level in human saliva due to oral rinsing [38], or the metabolic responses of yeast at different starving conditions [39].

2.1.2. Clustering

Clustering is a data analysis technique used to regroup unlabeled data on the basis of their similarities or differences. Examples of clustering algorithms are essentially the following: exclusive, overlapping, hierarchical, and probabilistic clustering [40,41]. Exclusive and overlapping clustering can be described together because they differ for the existence of one or multiple data points in one or more clustered sets. In fact, while exclusive clustering establishes that a data point can occur only in one cluster, overlapping clustering enables data points to be part of multiple clusters with different degrees of membership. Exclusive and overlapping clustering are hard or k-means clustering and soft or fuzzy k-means clustering, respectively [42–44]. In hard clustering, every element in a database might be a part of a single and precise cluster, whereas in soft clustering, there is a probability of having each data point into a different cluster [44]. Generally speaking, k-means clustering is a "distance-based" method in which each "clustered set" is linked with a centroid that is considered to minimize the sum of the distances between data points in the cluster.

Hierarchical clustering analysis (HCA) is used to recognize non-linear evolution in the data—contrary to what was done by the PCA which shows a linear trend—by means of a regrouping of features sample by sample without having any previous information [45]. This clustering method could be divided in two groups: (i) agglomerative clustering, and (ii) divisive clustering [46,47]. The first one allows to keep data points separate at first, unifying them iteratively later until it one cluster with a precise similarity between the data points is obtained. In the opposite way, divisive clustering creates a separation of data points in a data cluster on the basis of their differences. The clustering analysis leads to dendrograms that are diagrams in which the horizontal row represents the linked residues, whereas the vertical axis describes the correlation between a residue and previous groups. Figure 3 reports a dendrogram obtained by means of hierarchical cluster analysis performed on ^{1}H NMR data on the plasma metabolome of 50 patients with early breast cancer [48]. This kind of analysis allowed to discriminate among three different groups: LR-1 (red), LR-2 (blue) and LR-3 (green). They are characterized by significantly different levels of some metabolites, such as lactate, pyruvate and glutamin [48]. Furthermore, covariance analysis of NMR chemical shift changes allows defining functional clusters of coupled residues [49].

Clustering has been largely applied for metabolomic studies covering fields from medicine to food science, as is reported in the Applications section (Section 4). Here, we anticipate that clustering is essentially adopted for samples' classification by grouping metabolites without any external bias. This allows entering into the details of the precise metabolic pathways that may provide a connection between metabolomics and molecular biology. In such a way, many biomedical applications, including diagnostics and drug synthesis, would reach important improvements.

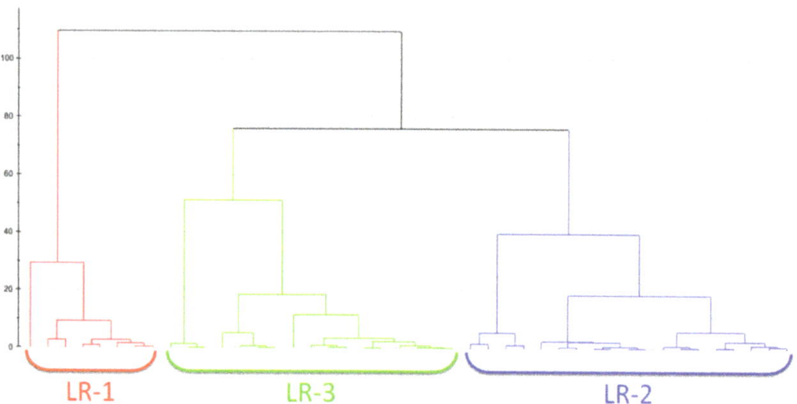

Figure 3. An example of a dendrogram obtained by means of hierarchical cluster analysis performed on ^{1}H NMR data on the plasma metabolome of 50 patients with early breast cancer. From the analysis, 3 different groups are classified: LR-1 (red), LR-2 (blue) and LR-3 (green). In this case, the Ward algorithm is adopted for measuring the distance. Figure reprinted from Ref. [48] under the terms of the CC-BY license.

2.1.3. Self-Organizing Maps (SOMs)

Self-organizing maps (SOMs) were introduced by Kohonen [50] and are widely employed to cluster a database, reduce its dimension and detect its properties by projecting the original data in a new discrete organization of smaller dimensions. This is performed by weighting the data throughout proper vectors in order to achieve the best representation of the sample. Starting from a randomly selected vector, the algorithm constructs the map of weight vectors for defining the optimal weights, providing the best similarity to the chosen random vector. Vectors with weights close to the optimum are linked with each unit of the map allowing to categorize objects in map units. Then, the relative weight and the total amount of neighbors reduce over time. Therefore, SOMs have the great power of reducing the dimensionality of the system while preserving its topology. For that reason, they are commonly adopted for data clustering and as a visualization tool. Another great asset of SOMs concerns the shapes of the clusters that do not require being chosen before applying the algorithm, whereas other clustering techniques usually work well on specific cluster shapes [51]. However, some limitations are evidenced using SOMs. In fact, they are normally of low quality, and the algorithm must be run many times before a satisfactory outcome is reached. Further, it is not easy to furnish information about the whole data distribution by only observing the raw map. Figure 4 reports the cluster of subjects involved in the study of renal cell carcinoma (RCC) by (NMR)-based serum metabolomics that was generated by using SOM (including the weighted maps for the considered 16 metabolites) [52].

The achieved results clearly separate healthy subjects (left region) and RCC patients (right region) within the SOM. Moreover, the weighted maps of the individual metabolites allow to identify a biomarker cluster including the following seven metabolites: alanine, creatine, choline, isoleucine, lactate, leucine, and valine. These may be considered for an early diagnosis of renal cell carcinoma [52].

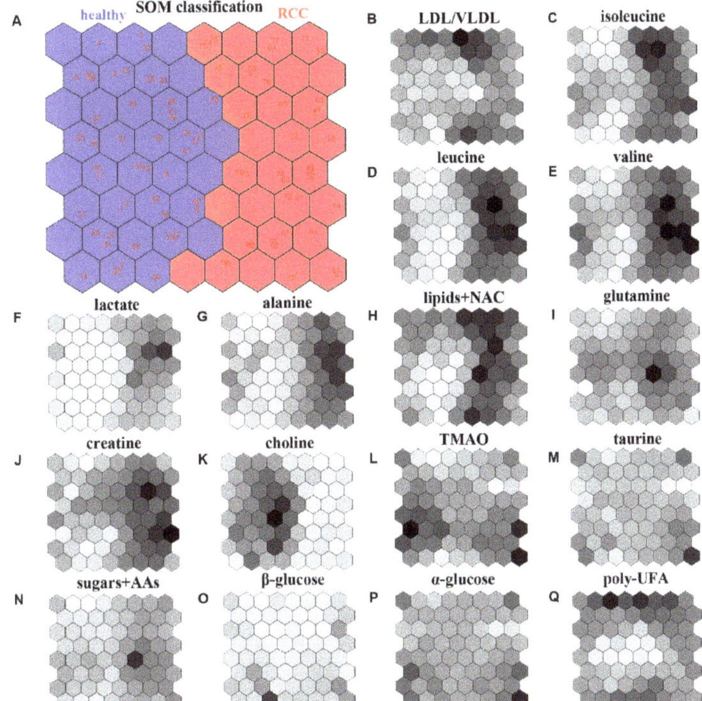

Figure 4. An example of SOM model for studying renal cell carcinoma (RCC). (**A**) SOM classification and discrimination between healthy subjects (left region) and RCC patients (right region) by considering 16 metabolites extracted by means of NMR spectroscopy on serum samples. (**B–Q**) Weight maps of the considered 16 metabolites. Darker colors correspond to higher SOM weights. Figure reprinted from Ref. [52] under the terms of the CC-BY license.

2.2. Supervised Methods

Problems or datasets having response variables (discrete or continuous) are generally treated with supervised methods. We distinguish between classification or regression problems, depending on whether the variables are discrete or continuous, respectively. The supervised technique is based on the association between the response variable (used to drive the model training) and the predictors (namely covariates) with the aim to perform precise predictions [53–55]. In fact, first, a training dataset is used as fitting model, while, in a second step, a testing dataset is used to estimate the predictive power. The relevant predictors are chosen by three types of feature selection methods [56] whose merits and demerits are listed in the scheme drawn in Figure 5 [57]:

1. The filter method marks subgroups of variables by calculate "easy to compute" quantities ahead of the model training.
2. The wrapper method marks subgroups of variables by applying the chosen trained models on the testing dataset with the aim to determine the achieving the optimal performance.
3. The embedded method is able to ascertain simultaneously the feature selection and model structure.

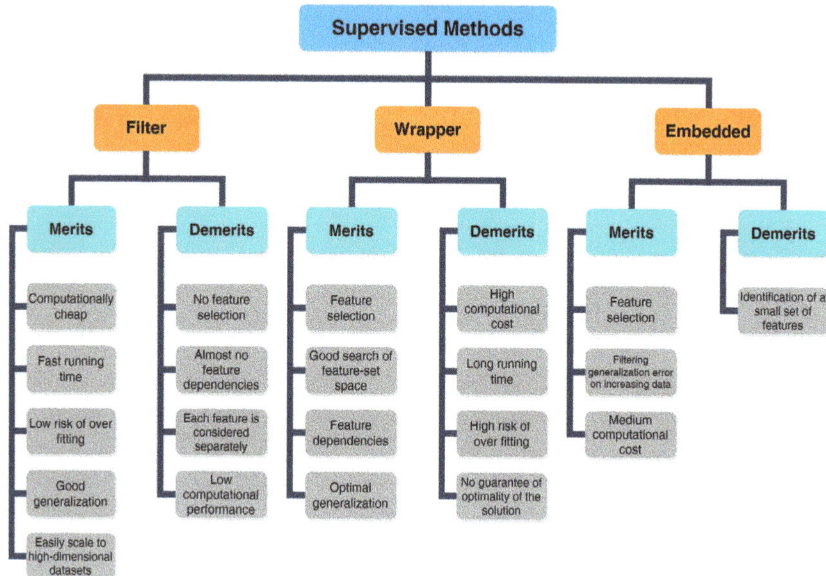

Figure 5. Scheme about merits and demerits of supervised methods, including filter, wrapper and embedded feature selection approaches.

Then, to measure the robustness of the fitting model and the predictive power, statistical approaches are adopted. Among them, we mention the root mean square error for calculating regression, sensitivity and specificity and the area under the curve for achieving classification.

For simplicity, let us consider that in binary classification, the test prediction can provide the following four results: true positive (TP), false positive (FP), true negative (TN), and false negative (FN). The model sensitivity, which coincides with the TP rate (TPR, i.e., the probability of classifying a real positive case as positive), is defined as TP/(TP + FN). On the contrary, the specificity is defined as TN/(FP + TN) and is linked to the ability of the test to correctly rule out the FP (FP rate, FPR = 1 − specificity). In order to evaluate the performance of binary classification algorithms, the most used approach is that of the receiver operating characteristic (ROC) curve, which consists of plotting TPR vs. FPR for the considered classifier at different threshold values (see Figure 6). The performance of the classifier is usually indicated by the value of the corresponding area under the ROC curve (AUC). Figure 6 shows, as an example, the ROC curve and the corresponding AUC value for a classifier with no predicting power (red dashed line with AUC = 0.5), a perfect classifier (green dotted line with AUC = 1) and a classifier with some predictive power (blue solid line with AUC$\sim$0.8).

Furthermore, several resampling methods, including bootstrapping and cross validation, can be adopted to achieve more reliable outcomes. This is a general description of the supervised methods; in the next, we will briefly enter into the details for some of them including random forest (RF) and k-nearest neighbors (KNN), principal component regression (PCR), partial least squares (PLS), and support vector machine (SVM).

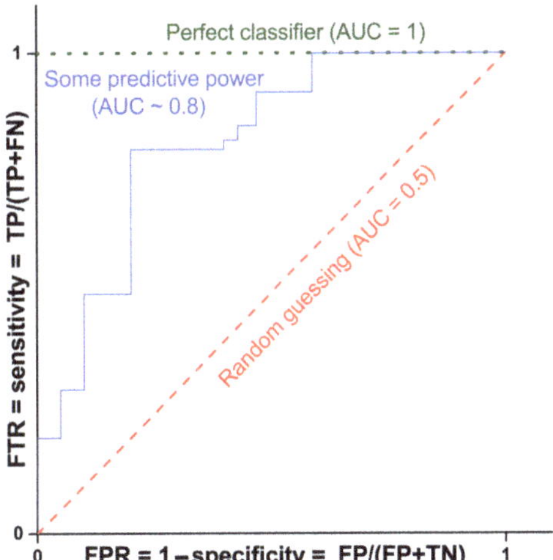

Figure 6. ROC curves and corresponding AUC values for three classifiers: no predicting power (red dashed line with AUC = 0.5), perfect classifier (green dotted line with AUC = 1) and some predictive power (blue solid line with AUC∼0.8).

2.2.1. Random Forest (RF) and k-Nearest Neighbors (KNN)

Although RF and KNN algorithms can be used for both supervised and unsupervised statistical analyses, here, we deal with the supervised aspects.

Random forest, as the name itself suggests, is composed by a proper number of decision trees working as an ensemble but individually depict a class from which the most representative corresponds to the model's prediction. Therefore, the idea behind the random forest algorithm is to correct the error obtained in one selection tree by using the predictions of many independent trees and by using the average value predicted by all these trees [58]. RF can deal with categorical features by treating both high dimensional spaces and a large number of training examples. In detail, the first step in a RF scheme is to create a selection tree; then, by using the observations $\{Y_j, X_j\}_{1<j<K}$, where X_j is usually a vector and Y_j is a real number, different sets can be obtained using different splitting criteria which operate on the considered vectors. Each criterion allows the initial subset to be divided into two subsets. An example of the selection tree is shown in Figure 7:

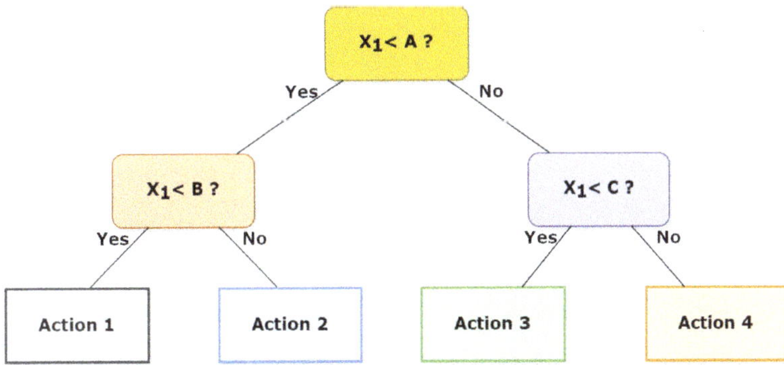

Figure 7. Example of decision tree with a different action corresponding to a different conditions set.

Given an observation X_i, and known selection tree, one determines in which final node the vector X_i is classified in order to predict Y.

Instead, the k-nearest neighbors (KNN) algorithm considers that similar outcomes lie near each other. Given again an observation X_i and with the aim to predict Y, the KNN algorithm selects the k-nearest observations of X_i in $\{Y_j, X_j\}_{1<j<K}$. Let $i_1, \ldots, i_k$ be the k values which provide the k minimum values of the function: g(j) = d(X_j - X_i). These minimum values can be equal if there are multiple values of X_j at the same distance from X_i [59]. There are at least the three possibilities for the distance (Euclidean, Manhattan and Minkowski). So, the value predicted for Y_i is the mean value of the k values Y_j for the k nearest neighbors of X_i:

$$\hat{Y}_i = \frac{1}{k}\sum_{1}^{k} Y_{ik} \tag{3}$$

2.2.2. Principal Component Regression (PCR) and Partial Least Squares (PLS)

It is well known that a linear model can be written as $Y = X\beta + \epsilon$, in which Y represents the response variable (it can be a single variable or even a matrix), and X represents the design matrix having variables along its columns and observations along its rows; β corresponds to the coefficients vector (or matrix) and ϵ represents the random error vector (or matrix). For a small number of variables and a high number of observations, it is commonly adopted for β the ordinary least square solution ($(X^TX)^{-1}X^TY$). In the opposite case, where it is not possible to evaluate the inverse of the singular matrix (X^TX), other solutions have to be considered [60]. One of them is the principal component regression (PCR) that makes use of the first PCs achieved by running PCA to fit the linear regression model instead of using all original variables. However, often, there is not a good correlation between these PCs and the response variables Y. Alternatively, the partial least squares (PLS) regression method is more efficient [61]. In the latter case, one has to determine the most suitable number of components to maintain, and then PLS evaluates a linear regression model by employing the projection of predicted and observed variables to a new space according to the following relations:

$$Y = UQ' + F \tag{4}$$

$$X = TP' + E \tag{5}$$

where T and U, analogously to PCA, correspond to X and Y scores and are matrices constituted by latent variables; at the same time, P' and Q' correspond to X and Y loadings, representing the weight matrices of the linear combinations; E and F represent all that is not possible to explain by using latent variables. Each of them, being expressed as a linear combination of X and Y, can be rewritten in terms of weight factors as $t = Xw$ and $u = Yc$, where t and u are two latent variables and w and c are the corresponding weight vectors. Indeed, PLS evaluates that set of X variables that is able to explain the majority of the changes in Y variables. Therefore, PLS, by using orthogonal conditions, evaluates those latent variables t and u, whose covariance is maximal. Ultimately, there are some substantial differences between the PCA and PCR-PLS approaches. In fact, as already mentioned, PCA pertains to unsupervised methods, whereas PCR and PLS pertain to supervised approaches. Moreover, as already mentioned, PCR takes advantage of the first PCs obtained from the PCA, using them as predictors for fitting the regression of a latent variable. Hence, PCA is able to explain just the X variance, whereas PLS allows achieving a multi-dimensional route in the X space, indicating the maximum variance route in the Y space. In other words, in PCR, the principal components become the new (unrelated) variables of the regression, which thus becomes more easily resolvable. Otherwise, in PLS, the Y variables are decomposed into principal components too, while those of X are rotated along the direction of maximum correlation with respect to the principal components of Y. Therefore, the purpose of PLS is to determine latent variables similar to the principal components that maximize the variance of both matrices.

We also mention the partial least squares discriminant analysis, or PLS-DA, which is an alternative when the dependent variables are categorical. Discriminant analysis is a classification algorithm which adds the dimension reduction part to it. PLS-DA allows the employment of predictive and descriptive algorithms other than for discriminative variable choice (see Figure 8a). PLS-DA is executed on NMR spectra for different aims, including food authentication and diseases classification in medical diagnostics [62–65]. However, a more comprehensive variant of PLS is the orthogonal PLS (OPLS) method. It is finalized to separate systematic changes in X into two parts; one of them is in linear relationship with Y and another is irrelevant to Y (generally, perpendicular to it). So, some changes in X which are perpendicular to Y are eliminated, while uncorrelated changes in X are separated from correlated ones (see Figure 8b). In this way, the uncorrelated changes are analyzed separately, favoring the prediction ability and the interpretation of results [66]. This latter is one of the advantage of OPLS with respect to PLS together with the aspect that the inner repetition is not time consuming, which can accelerate the calculation process. In fact, OPLS is more appropriate for discriminating the precise differences between two systems, providing information on the variables with the largest discriminatory power.

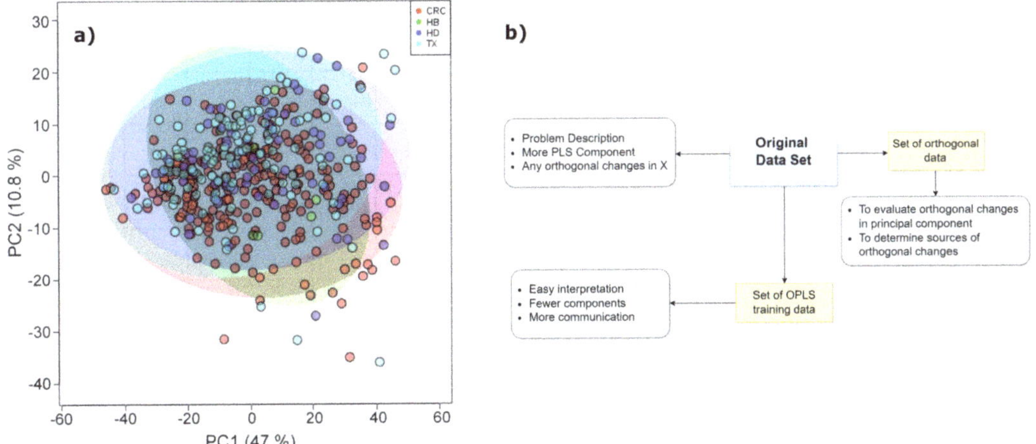

Figure 8. (**a**) Bidimensional PLS-DA score plot of urine samples obtained from different hospitals. HB—Basurto Hospital, CRC—Cruces Hospital, HD—Donosti Hospital, TX—Txagorritxu Hospital. Figure reprinted from [67] under the terms of Creative Commons Attribution 4.0 International License. (**b**) OPLS scheme.

2.2.3. Support Vector Machine (SVM)

Considering the data organized into a matrix, each subject corresponding to a row vector can be conceived as a single point in the p-space of the considered variables. Data can be essentially organized into two main groups, "separated by a gap" whose margins are defined by support vectors. Instead, the edge located in the gap center separating the data corresponds to the dividing hyperplane. SVM tries to define the support vectors, and the prediction will indicate to which hyperplane side the new observations belong. However, generally, data cannot be linearly separated, and hence it is difficult to determine the separating hyperplane. Nevertheless, SVM can accurately execute a non-linear classification throughout the so-called kernel trick. It consists of an implicit mapping of the considered inputs into high-dimensional feature spaces with the objective to their linear separation in that space [68]. In detail, the optimal hyperplane is the one that provides the highest separation between the two classes. With greater definition, by separation, we mean the maximum amplitude (or width, w) between the lines parallel to the hyperplane without any data points in between. This optimal hyperplane is called the maximum-margin hyperplane and the corresponding linear classifier is the maximum-margin classifier (Figure 9).

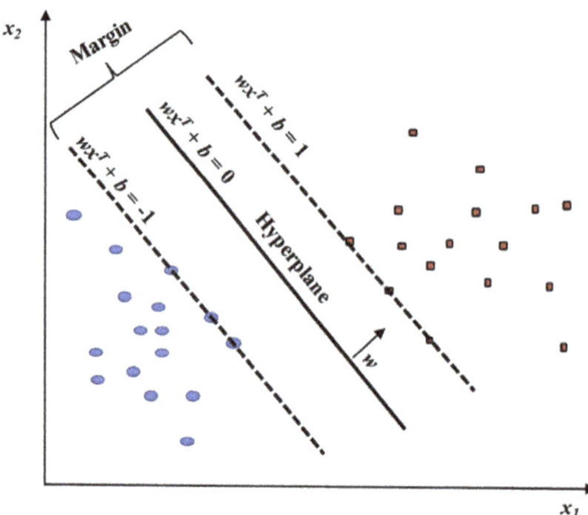

Figure 9. Linear SVM model highlighting the classification of two classes (red and blue). Figure reprinted from Ref. [68] under the terms of the HighWire Press license.

In addition, in the presence of mislabeled data, the SVM can provide inadequate classifications; therefore, only a few misclassified subjects are found instead by maximizing the separation between the two classes. Finally, validation methods and diagnostic measures are analogous to those adopted in PLS methods. Ultimately, SVM is one of the approaches with the highest accurate prediction, since it is based on statistical learning frameworks [69,70]. It can also be used within machine learning approaches for anomaly detection (such as weather) by choosing an anomaly threshold with the aim to establish whether an observation belongs to the "normal" class or not. Disadvantages of supervised methods include overfitting problems [71] corresponding to the inclusion of noise inside the statistical model. These issues can be provoked by excessive learning, so several validation techniques, such as cross validation [72] or bootstrapping [73], are usually employed to solve them.

2.3. Pathway Analysis Methods

A powerful method to describe peculiar features of the cell metabolism is pathway analysis (PA), which provides a graphical representation of the relationships among the actors (mainly enzymes and metabolites) of precise catalyzed reactions. Therefore, PA is highly employed for the interpretation of high-dimensional molecular data [74]. In fact, taking advantage of the already acquired knowledge of biological pathways, proteins, metabolites and also genes can be mapped onto newly developed pathways with the objective to draw their collective functions and interactions in that specific biological environment [75]. Although PA was initially developed for the interpretation of transcriptomic data, in the last decades, it has become a common method in metabolomics, being particularly suited to find associations between molecules involved in the same biological function for a given phenotype [76–78].

PA methods include several tools allowing deep statistical analyses in metabolomics known as enrichment analysis. They grant the functional interpretation of the achieved results mainly in terms of statistically significant pathways [79]. These tools can handle heterogeneous and hierarchical vocabularies and may be classified into two distinct collections. The first encompasses "non-topology-based" (non-TB) approaches, which do not consider the acquired knowledge concerning the character of each metabolite in the considered pathways [80]. Non-TB approaches include the over-representation analysis (ORA) as the first generation technique and the functional class scoring (FCS) as the second generation.

Finally, the second collection includes topology-based methods (see Figure 10) that are adopted to determine those pathways that are significantly impacted in a given phenotype.

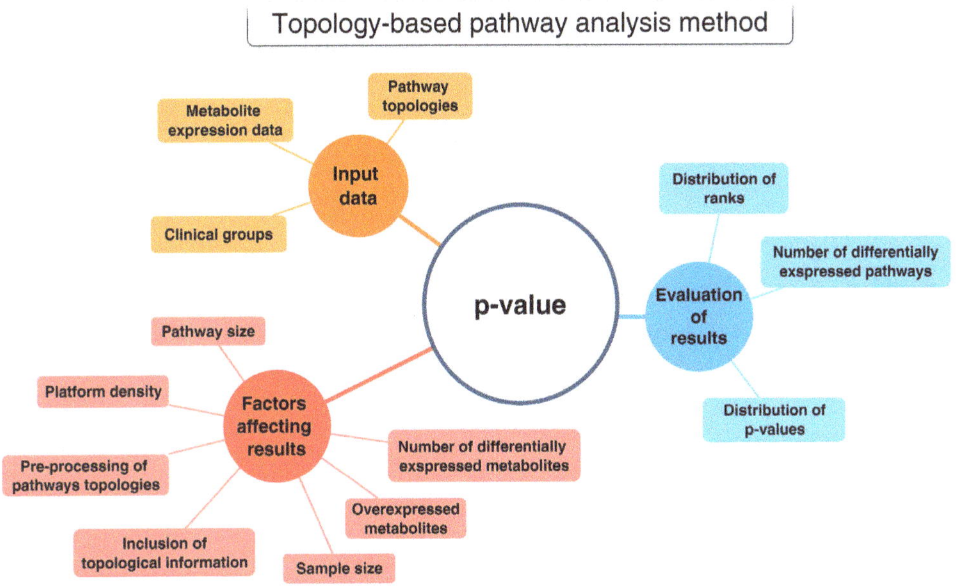

Figure 10. Conceptual map about the topology-based pathway analysis method.

This latter approach can be classified depending on the considered pathways (e.g., signaling or metabolic), inputs (e.g., subset or all metabolites and metabolites *p*-values), chosen mathematical models, outputs (e.g., pathway scores and *p*-values) and the wanted implementation (e.g., web-based or standalone) [81,82]. Note that PA methods were originally developed for genes, but they can be successfully applied for every biomolecule/metabolite [83].

2.3.1. Over-Representation Analysis (ORA)

Over-representation analysis (ORA) is among the most used pathway analysis approaches for the interpretation of metabolomics data needed as input, once the type of annotations to examine is chosen. One obtains a collection of annotations and their associated *p*-value as outputs since a statistical test is applied to determine whether a set of metabolites is enriched by a specific annotation (e.g., a pathway) in comparison to a background set. Different statistics can be applied to obtain information about the studied biological mechanisms and on the specific functionality of a given metabolite set. Among the most used statistics, we would like to mention the well-known binomial probability, Fisher's exact test and the hypergeometric distribution [84,85].

Three are the necessary inputs in ORA analysis: (i) a set of pathways (or metabolite collections); (ii) a catalog of investigating metabolites and, (iii) a background collection of compounds. The list of investigating metabolites usually comes from experimental data after applying a statistical test to determine those metabolites whose signals can be associated with a precise result by choosing a threshold value usually associated to the *p*-values [74]. The background collection includes all metabolites that can be revealed in the considered measurement. If the p-value corresponding to each pathway is obtained by means of the right-tailed Fisher's exact test based on the hypergeometric distribution, the probability to find k metabolites or more in a pathway can be written as [74]:

$$P(X \geq k) = 1 - \sum_{i=0}^{k-1} \frac{\binom{M}{i}\binom{N-M}{n-i}}{\binom{N}{n}}, \tag{6}$$

where N corresponds to the number of background compounds, n is the number of the measured metabolites, M is the number of background metabolites mapping the ith pathway, and k represents the overlap between M and n. A scheme of the ORA principle is displayed in Figure 11 as a 3D Venn diagram. Finally, multiple corrections are usually applied, as calculations are made for many pathways, thus obtaining a collection of significantly enriched pathways (SEP).

Figure 11. A 3D Venn diagram illustrating the relation between ORA parameters (Equation (6)) in which N corresponds to the number of background compounds, n is the number of the measured metabolites, M is the number of background metabolites mapping the ith pathway, and k represents the overlap between M and n.

Before applying ORA, one has to verify if the metabolomics dataset is sufficiently big to furnish proper statistical significance. For instance, usually MS-based techniques can observe more metabolites than NMR-based methods, such as the mono-dimensional NMR ones commonly used for profiling [86]. Indeed, the choice of the most suitable background collection is the real challenge and still remains an open subject because it strictly depends on the situation [74].

2.3.2. Functional Class Scoring (FCS)

Functional class scoring (FCS) methods look for coordinated variations in the metabolites belonging to a specific pathway. In fact, FCS methods take into account those coordinated changes within the individual set of metabolites that, although weak, can have a significant effect of specific pathways [75,78]. Essentially, all FCS methods comprise three steps (see Figure 12):

1. A statistical approach is applied to compute differential expression of individual metabolites (metabolite-level statistics), looking for correlations of molecular measurements with phenotype [87]. Those mostly used consider the analysis of variance (ANOVA) [88], Q-statistic [89], signal-to-noise ratio [90], t-test [91], and Z-score [92]. The choice of the most suitable statistical approach may depend on the number of biological replicates and on the effect of the metabolites set on a specific pathway [93].
2. Initial statistics for all metabolites of a given pathway are combined into statistics on different pathways (pathway-level statistics) that can consider interdependencies among metabolites (multivariate) [94] or not (univariate) [91]. The pathway-level statistics usually is performed in terms of the Kolmogorov–Smirnov statistics [90], mean or median of metabolite-level statistics [93], the Wilcoxon rank sum [95], and the maxmean statistics [96]. Note that, although multivariate statistics should have more statistical significance, univariate statistics provide the best results if applied to the data of biologic systems ($p \leq 0.001$) [97].

3. The last FCS step corresponds to estimating the significance of the so-called pathway-level statistics. In detail, the null hypothesis can be tested into two different ways: (i) by permuting metabolite labels for every pathways, so comparing the set of metabolites in that pathway with a set of metabolites not included in that pathway (competitive null hypothesis) [75] and (ii) by permuting class labels for every sample, so comparing the collection of metabolites in a considered pathway with itself, whereas the metabolites excluded by that pathway are not considered (self-contained null hypothesis) [91].

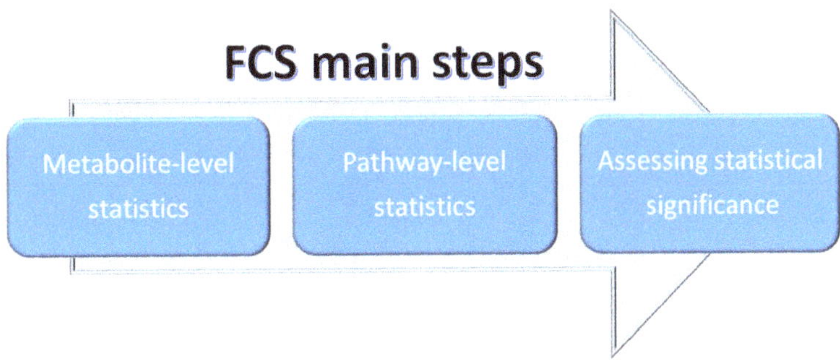

Figure 12. Schematic representation of the three main steps adopted in FCS methods.

2.3.3. Metabolic Pathway Reconstruction and Simulation

The identification of metabolomic biomarkers and their mapping into a neural network is fundamental to further study the cellular mechanisms and its physiology. The goal is to identify the effects of the metabolites (as a function of their concentration) on the cellular changes, providing a relationship with the most likely biologically meaningful sub-networks. Thus, basing on genome annotation and protein homology, reference pathways could be mapped into a specific organism. However, this mapping method often produces incomplete pathways that need the employment of ab initio metabolomic network construction approaches (such as Bayesian networks), where differential equations describe the changes in a metabolomic network in terms of chemical amounts [98,99]. Qi et al. [100] further improved this approach allowing to optimize accuracy in defining metabolomics features or better the correlation between the substrates whose nature is well known as well as the species of each individual reactions, so defining the classification of the mapped metabolic products in a pathway and their modifications under selected perturbations. Recently, Hu et al. [23] performed a pathway analysis on serum spectra recorded by ^{1}H NMR with the aim to identify eventual biomarkers characterizing the treatment of human lung cancer. After a first statistical analysis in terms of PLS-DA, they were able to identify four metabolic pathways associated with the metabolic perturbation induced by non-small-cell lung cancer (Figure 13) by means of the MetaboAnalyst package [101]. In detail, the highest pathway impact was shown by the metabolisms of (i) taurine and hypotaurine, (ii) d-glutamine and d-glutamate, (iii) glycine, serine and threonine, and (iv) alanine, aspartate and glutamate, thus shedding light on the responsible processes in this kind of cancer.

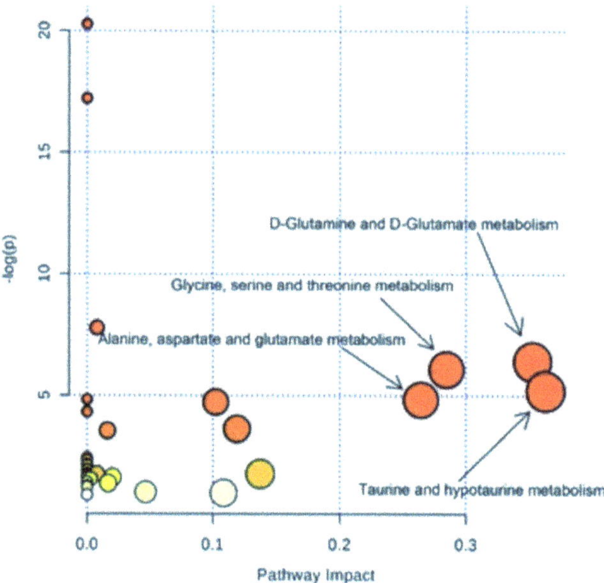

Figure 13. Pathway analysis performed on serum spectra recorded by ^{1}H NMR allowing the identification of main metabolic pathways associated with non-small cell lung cancer. The larger the circle, the higher the impact. The color, from red to yellow, identifies the corresponding significance. Figure reprinted from [23] under the terms of the Creative Commons Attribution 4.0 International License.

3. Artificial Intelligence toward Learning Techniques

Artificial intelligence (AI) techniques are based on algorithms that try to simulate both human learning and decision making. Indeed, AI exploits the ability of computer algorithms to learn from a given dataset containing precise information that then must be recognized in new dataset in an automatic way. Specifically, the computer algorithms during learning on the test dataset create models that are able to provide information on the probability that a specific result may occur. Furthermore, these programs are usually able to identify the important features associated with the outcome of interest. Artificial intelligence methods can accurately handle big data for biomarkers prediction, allowing the determination of relevant characteristics pertaining to a dataset and a deep comprehension of the significance of such data. Specifically, the integration of metabolic snapshots with metabolic fluxes and the use of knowledge-informed AI methods allow obtaining a profound comprehension of metabolic pathways at the system level. Hence, the development of multi-omic techniques integrating both experimental and computational methods, adequate to extract metabolic information at the cellular and subcellular levels, will provide powerful tools to enter the details of metabolic (dis)regulation, therefore allowing the exploitation of personalized therapies [102].

Machine Learning, Neural Networks and Deep Learning

All the conventional approaches discussed in the previous sections can be implemented by learning algorithms that let the corresponding network learn by a given dataset and, after performing a test with a sample dataset, can be used with a known predictive power. In this section, we get into details of the different machine learning techniques as a subset of AI methods (Figure 14).

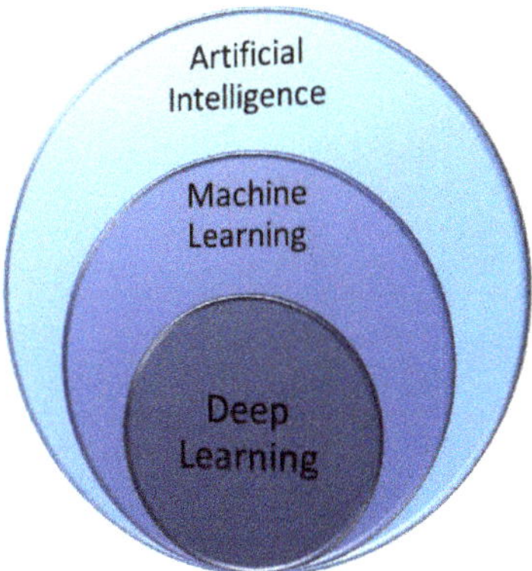

Figure 14. Venn diagram illustrating that deep learning is the core of machine learning, which in turn is a technique within AI methods.

In addition, neural networks and deep learning approaches are characterized in terms of the number of node layers, also named depth. Briefly, a node is the locus in which the algorithm performs the calculation and would correspond to the action that a neuron exerts in the human brain when it is subject to a stimulus. As shown in Figure 15b, a node takes different inputs, each having its own weight, that can be amplified or reduced by the activation function, thus giving a corresponding significance to the received inputs with respect to the specific task that the used algorithm is learning. So, a neural network consisting of two or more hidden layers can be classified as a deep learning technique and is usually described by the diagram shown in Figure 15a, together with a scheme of how one node might look (Figure 15b).

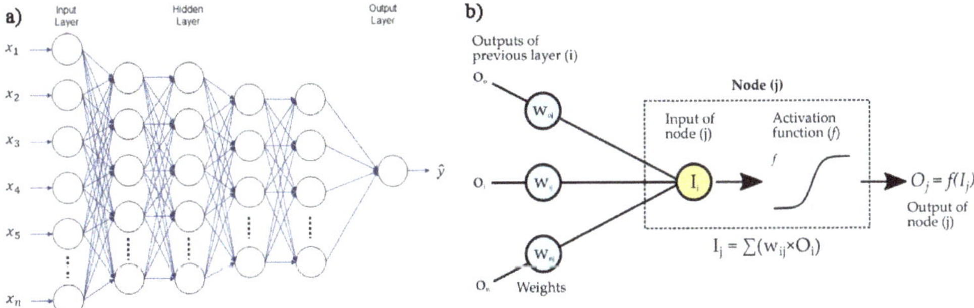

Figure 15. (**a**) Example scheme of a deep neural networks, reprinted from Ref. [103] under the terms of the CC-BY license; (**b**) operating principle of a single node.

Deep learning techniques, being able to handle large datasets, thus allowing a high-level description, are already used to provide the optimal route to solve a lot of issues in the field of image recognition, speech recognition, and natural language processing. Furthermore, DL techniques can be divided into three main categories (see Figure 16) that are deepened in Ref. [104]:

- Supervised learning (discriminative) includes multi-layer perceptron (MLP), convolutional neural network (CNN), long short-term memory (LSTM) and gated recurrent unit (GRU);
- Unsupervised learning (generative) includes generative adversarial network (GAN), autoencoder (AE), sparse autoencoder (SAE), denoising autoencoder (DAE), contrac-

tive autoencoder (CAE), variational autoencoder (VAE), self-organizing map (SOM), restricted Boltzmann machine (RBM) and deep belief network (DBN);

- Hybrid learning (both discriminative and generative) includes models composed by both supervised and unsupervised algorithms other than deep transfer learning (DTL) and deep reinforcement learning (DRL).

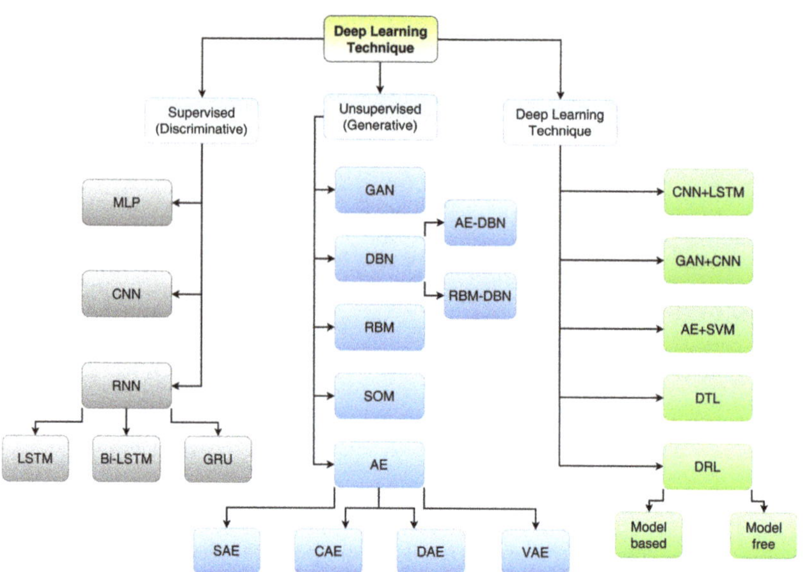

Figure 16. A taxonomy of DL techniques. For acronyms, see main text.

Supervised learning can furnish a discriminative function in classification applications by identifying the different features of those classes that can be extracted by the data. Among them, multi-layer perceptron (MLP) is a feedforward ANN that involves (i) an input layer collecting input signals, (ii) an output layer that provides an outcome in consideration of the processed input and (iii) some hidden layers separating the input and output layers that correspond to the network computational engine. On the contrary, unsupervised learning is employed to recognize eventual correlations by analyzing the signals pattern and to assess the statistical distributions of the achieved results both on original data and on their corresponding classes. This kind of generative approach can be also used as an initial step (pre-processing) before applying supervised learning methods. Most common unsupervised techniques, reported in Figure 16, are listed and briefly described in the next. Hybrid learning paradigms combining both discriminative and generative methods are possible. Hybrid deep learning architectures are usually constituted by multiple models where the basis can indeed be either a supervised or unsupervised deep learning method. Common hybrid learning algorithms are, for example, semi-supervised learning that allows to use a supervision for some data points, keeping the others unlabeled, and deep reinforcement learning (DRL; see Figure 16) that, interacting with an environment, involves the knowledge of performing with sequential decision-making tasks in order to maximize cumulative rewards [104,105]. Their advantages lie in the possibility to consider the best aspects of discriminative and generative models. For instance, a hybrid architecture can adopt small inputs to avoid the problem of determining the right network size and instead an increasing number of neurons in receptive-field spaces [106]. At the same time, by a proper enhancement of the initial weights through suitable algorithms, neural networks in hybrid architectures can provide higher accuracy and predictive power [107,108].

Most of the techniques in the categories indicated before are feed-forward (working from input to output) but, as detailed in the last part of the section, the opposite movement is also possible. This is called backpropagation and works from output to input, allowing the evaluation of the individual neuron's error, allowing to properly modify and fit the algorithm iteratively. Unlike ML, that usually adopts manual identification and description of relevant features, DL techniques aim to execute automatically the features extraction, avoiding almost all human participation. In addition, DL can handle larger datasets, especially of the unstructured type. In fact, DL methods can have unstructured raw data as input (such as text or images) and can directly define which characteristics must be considered to distinguish the original observations. By recognizing similar and/or different patterns, DL methods can adequately cluster inputs. Therefore, DL approaches would need a very high number of observations to be as accurate as possible. Generally, and according to the scheme reported in Figure 16, the most adopted deep learning techniques are the following [104]:

1. **Classic neural networks** encompass linear and non-linear functions which, in turn, include S-shaped functions ranging from 0 to 1 (sigmoid) or from −1 to 1 (hyperbolic tangent, tanh) and rectified linear unit (ReLU), which gives 0 for input lower than the set value or evaluates a linear multiple for bigger input.
2. **Convolutional neural networks (CNN)** take into high consideration the neuron organization found in the visual cortex of an animal brain. It is particularly suited for high complexity and allows for optimal pre-processing. Four stages can be considered for CNN building (see Figure 17):
    (a) Deduce feature maps from input after applying a proper function (convolution);
    (b) Reveal an image after given changes (max-pooling);
    (c) Flatten the data for the CNN analysis (flattening);
    (d) Compiling the loss function by a hidden layer (full connection).
3. **Recurrent neural networks (RNN)** are exploited when the objective is the prediction of a sequence. They are a subset of ANN for sequential or time series data, usually applied for language translation, speech recognition, and son on. Their peculiar feature is that the outcome of the output node is a function of the output of previous elements within the sequence (see Figure 18a).
4. **Generative adversarial networks (GAN)** combine generator networks for providing artificial data and discriminator networks for distinguishing real and fake data.
5. **Self-organizing maps (SOMs)** have a fixed bi-dimensional output since each synapse joins its input and output nodes, and usually take advantage of data reduction performed by unsupervised approaches.
6. **Boltzmann machine** is a stochastic model exploited for yielding proper parameters defined in the model.
7. **Deep reinforcement learning** are mainly used to understand and so predict the effect of every action executed in a defined state of the observation.
8. **Autoencoders** work directly on the considered inputs, without taking into account the effect of activation functions. Among the autoencoders, we mention the following:
    (a) Sparse autoencoders have more hidden than input layers for reducing overfitting.
    (b) Denoising autoencoders are able to reconstruct corrupted data by randomly assigning 0 to some inputs.
    (c) Contractive autoencoders include a penalty factor to the loss function to prevent overfitting and data repetition when the network has more hidden than input layers.
    (d) Stacked autoencoders perform two stages of encoding by the inclusion of an additional hidden layer.
9. **Backpropagation (BP)** are neural networks that use the flux of information going from the output to input for learning about the errors corresponding to the achieved prediction. An architecture of the BP network is shown in Figure 18b.

10. **Gradient descent** are neural networks that identify a slope corresponding to a relation among variables (for example, the error produced in the neural network and data parameter: small data changes provoke errors variations).

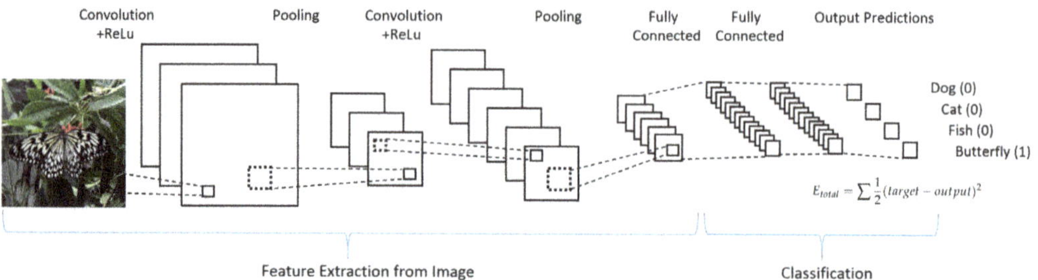

Figure 17. Example of a convolutional neural network. Figure reprinted from Ref. [109] under the terms of CC BY-NC-ND 4.0 license.

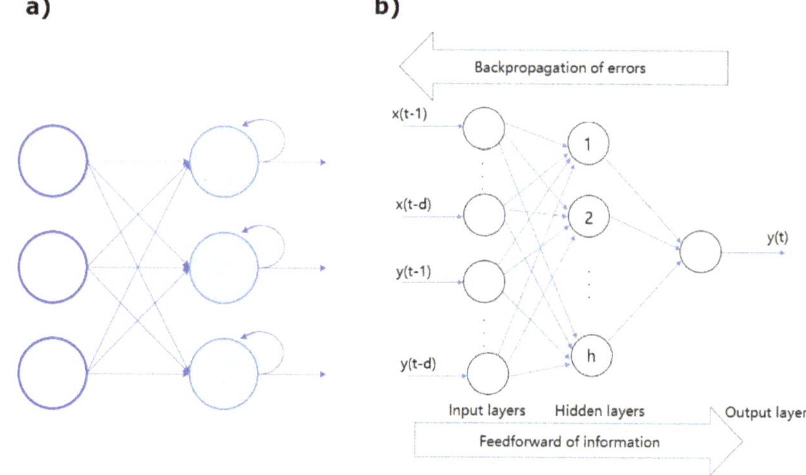

Figure 18. (**a**) Scheme of a RNN. (**b**) Example of a BP network architecture. Figure reprinted from Ref. [109] under the terms of CC BY-NC-ND 4.0 license.

From the above brief information, it emerges that, even if DL methods can be thought as black-box solutions, future generation deep learning can provide a great aid for the analysis of big data and for corresponding reliable results.

4. Applications of Deep Learning Approaches for NMR-Based Metabolomics

In this section, the applications of deep learning on NMR-based metabolic data for specific different fields are reported and discussed. Here, we briefly introduce the potentiality of the applications of deep learning in metabolomics which today are still relatively low compared to other omics. This is explained since metabolome-specific deep learning architectures should be defined, and dimensionality problems and model evaluation regime should be further evaluated. In any case, data pre-processing using convolutional neural network architecture appears to be the most efficient approach among the deep learning ones. The main advantage of CNNs compared to a traditional neural network is that they automatically detect important features without any human supervision. Specifically, CNNs learn relevant features from image/video at different levels, similar to a human brain [110]. This is very relevant to analyze both biomedical and food data, whose classification in view of safety security actions is extremely important.

The potentiality of the NMR technique within the field of metabolomics is currently employed for several purposes, including the detection of viable microbes in microbial food safety [10], the assessment of aquatic living organisms subjected to contaminated water [111], the identification of novel biomarkers to diagnose cancer diseases [112] and the monitoring of the plant growth status changing environmental parameters in view of smart agriculture [113]. In the next sections, we discuss some applications of deep learning approaches for NMR-based metabolomics in food and biomedical areas, highlighting their strengths and limitations.

Even before the development of artificial intelligence, statistical analyses were successfully applied in food analysis but with some limitations. For example, traditional methods are usually not very accurate in the classification of similar foods in contrast to modern deep learning approaches that allow enhancing all small differences. However, traditional methods usually constitute the first step, providing the input for neural networks with the aim to achieve a more accurate and automatic output. Furthermore, advanced computational algorithms can be applied not only for statistical analysis, but also to execute simulations whose predictions depend on the considered conditions [114].

4.1. Food

Foodomics is a term referred to the metabolomic approaches applied to foodstuffs for investigating topics mainly related with nutrition. Nowadays, DL methods are being progressively applied in the food field with different purposes, such as fraud detection [115]. Furthermore, another important issue is to guarantee the geographical origin and production/processing procedures of food, the precise proportions of ingredients, including additives and the kind of used raw materials. In this context, machine learning is a powerful method for achieving an adequate classification. For example, Greer et al. [116] carried out NMR measurements using a not-conventional protocol to measure the magnetization relaxation times (both the longitudinal T_1 and transverse T_2) and then they efficiently classified cooking oils, milk, and soy sauces (see Figure 19).

Since the considered datasets are very large (typically about 5×10^6 points each), the authors first reduced their size by means of the singular value decomposition, thus allowing a fast classification and also providing little insight into the sample physical properties. Figure 19 reports different combinations for the obtained classification features. Figure 19a,b corresponds to the two components used by the Gaussian fit of those peaks revealed by the inverse 2D Laplace transform [117]. A sharp distinction of the samples is clearly shown for every adopted combination. The y-axes of Figure 19a,c report the first component of T_1 versus the first and second components of T_2, respectively. Contrarily, the y-axes of Figure 19b,d report the second component of T_1 versus the first and second components of T_2, respectively. The authors found that most of the trained models reached an accuracy up to 100% (see, for example, Figure 20a). Finally, they also pointed out the effect of the sample temperature on classification accuracy for achieving reliable results (see Figure 20b).

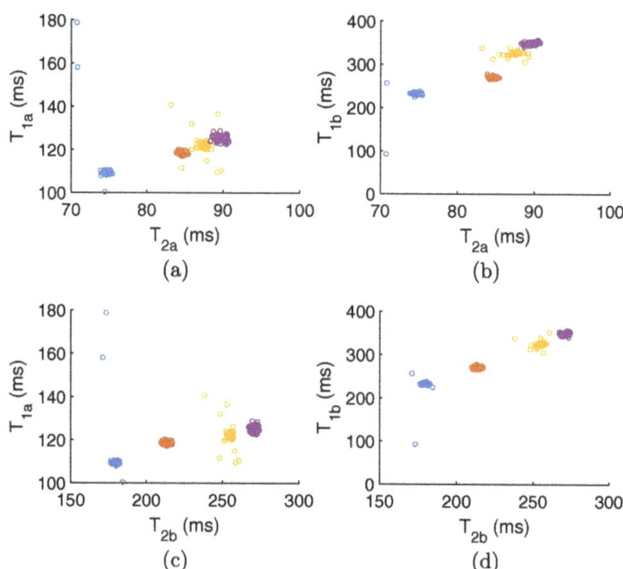

Figure 19. T_1–T_2 correlational maps classifying several kinds of oils: olive (blue), canola (orange), corn (yellow) and vegetable (purple) by using the two components used by the Gaussian fit of those peaks revealed by the inverse 2D Laplace transform. (**a**,**c**) report the first component of T_1 versus the first and second components of T_2, respectively. (**b**,**d**) report the second component of T_1 versus the first and second components of T_2, respectively. See main text and Ref. [116] for details. Figure reprinted with permission from Ref. [116]. Copyright 2018 Elsevier.

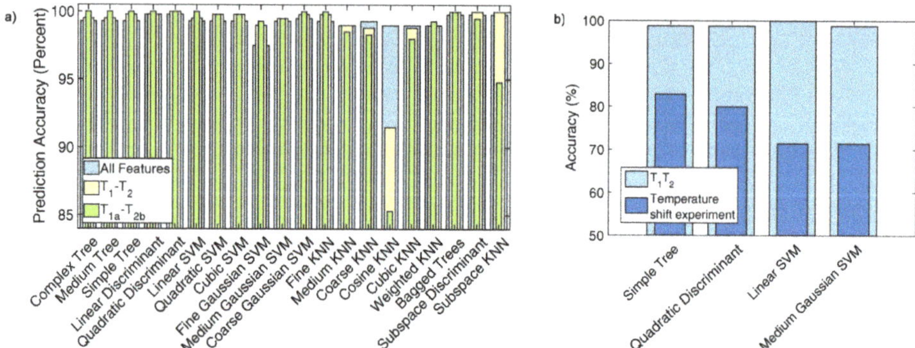

Figure 20. (**a**) Comparison of the accuracy for the predictive power of the algorithms applied to classify cooking oil samples by employing three different classification training; (**b**) accuracy of predictive power applied to soy sauce sample highlighting the effect of temperature. Figure adapted with permission from Ref. [116]. Copyright 2018 Elsevier.

Nowadays, deep neural networks (DNNs) are rarely used for metabolomics studies because the assignment of metabolites contribution in NMR spectra still lacks highly reliable yields due to the complexity of the investigated biological matrix and thus of the corresponding signals. As described in the previous section, different deep learning methods were used, but some of them are characterized by some limitations (i.e., low accuracy in classification). Some efforts were made to overcome this problem. Date et al. [118] recently developed a DNN method that includes the evaluation of the so-called mean decrease accuracy (MDA) to estimate every variable. It relies on a permutation algorithm that allows the recognition of the sample geographical origins and the identification of their biomarkers. On the other hand, for food authenticity and nutritional quality, the fast revelation of viable microbes is still a challenge. Here, we report a multilayer ANN example (see Figure 21) showing four input neurons, two hidden layers made of three neurons, and two output neurons.

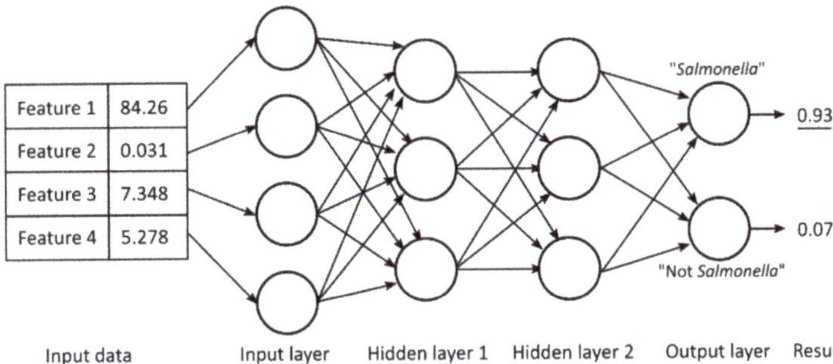

Figure 21. Multilayer artificial neural network showing 4 input neurons, 2 hidden layers made of 3 neurons, and 2 output neurons of which the one corresponding to "Salmonella" shows the highest value, associated with the prediction performed by the used ANN. Figure reprinted from Ref. [119] under the terms of the CC-BY license.

Such a scheme was organized by Wang et al. [119] for the detection, by means of NMR spectroscopy coupled with deep ANNs, of pathogenic and non-pathogenic microbes. According to the classification method, each output neuron is associated to one possible output. Here, "Salmonella" shows the highest value of output, thus corresponding to the prediction performed by the used ANN. In such a case, the weights of each input are optimized to reach the wanted outcome throughout backpropagation, thus defining multiple epochs and training cycles. Figure 22 reports an example referred to an ANN analysis with two hidden layers of 800 neurons. ANN training is made optimizing a set of training criteria to avoid shallow local minima. In particular, training continues when the loss function decreases after an epoch of training ("greedy" algorithm—case a) and even after a small increase followed by a continuous decrease (case b). On the contrary, training stops for an increase in the loss function after several constant values (case c) and for steep increases (case d) [119].

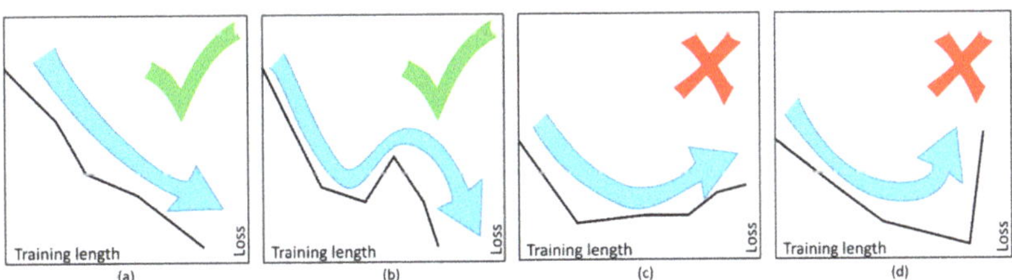

Figure 22. Comparison of the different criteria adopted for the ANN training. (**a**) "Greedy" learning; (**b**) "jumping" out of a local tiny minimum; (**c**) halt at large minima; (**d**) halt at sharp growths in loss. Figure reprinted from Ref. [119] under the terms of the CC-BY license.

Once the network is trained, it is able to perform predictions on new input data. As already mentioned, the loss and the model accuracy provide a measure of the output goodness. In fact, the aim is to minimize the disagreement between the prediction and the reality (loss) and to maximize accuracy (cross-validation method). Thanks to this approach, Wang et al. [119] found that the used ANNs accurately predict 91.2% of unknown microbes and, after repeating the model training by considering just those metabolites whose amount increased with incubation time, they observed an accuracy up to 99.2%.

Machine learning and neural network approaches are simultaneously adopted to analyze large amounts of NMR metabolomics data for food safety [109]. This can be performed also by means of magnetic resonance imaging (MRI), which is an imaging technique relying on NMR principles. Within the food field, it is mainly used to resolve the tissue texture of foods [120,121]. On the other hand, Teimouri et al. [122] used PLSR, LDA, and ANN for the classification of the data collected by CCD images from food portions, different in color and geometrical aspects. In this way, they were able to classify 2800 food samples in one hour, with an overall accuracy of 93%. Instead, De Sousa Ribeiro et al. [123] developed a CNN approach able to reconstruct degraded information on the label of food packaging. Before applying CNNs, they started with K-means clustering and KNN classification algorithms for the extraction of suitable centroids.

4.2. Biomedical

Metabolomics-based NMR investigations, coupled with deep learning methods, are increasingly employed within the biomedical field. More profoundly, the use of complex DL architectures hardly allows achieving a predictive power with ranking or selection. As already discussed, DL models use several computational layers to analyze input signals and establish any eventual preferred direction for signal encoding (forward or backward). This procedure does not usually allow the interpretation of input signals in terms of the used model, making it hard to identify biomarkers in a network, where biological and DL modeling are connected (Figure 23).

Today, it is still necessary to uniform assessment metric for biomedical data classification or prediction, also avoiding false negatives in disease diagnosis. Further, deep learning is a promising methodology to treat data collecting by smart wearable sensors, which is considered fundamental in epidemic prediction, disease prevention, and clinical decision making, thus allowing a significant improvement in the quality of life [124,125].

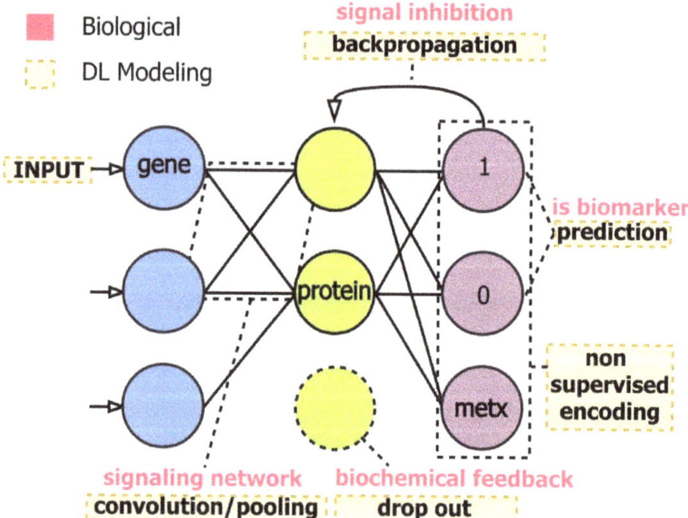

Figure 23. The multiomics method represented connects biological (i.e., signal inhibition, signaling network and biochemical feedback) with DL modeling (backpropagation, prediction, convolution, etc.), aiming to maximize the robustness of the approach for the identification of biochemical features referred to specific phenotypes. Figure reprinted from Ref. [124] under the terms of the Creative Commons Attribution Noncommercial License.

With the aim to obtain an accurate metabolites identification from the observation of the corresponding peaks in complex mixtures, Kim et al. [126] developed a convolutional neural network (CNN) model, called SMART-Miner, which is trained on 657 chemical entities collected from HMDB and BMRB databases. After training, the model is able to automatically carry out the recognition of metabolites from ^{1}H-^{13}C HSQC NMR spectra of complex metabolite mixtures, showing higher performance in comparison with other NMR-based metabolomic tools (Figure 24).

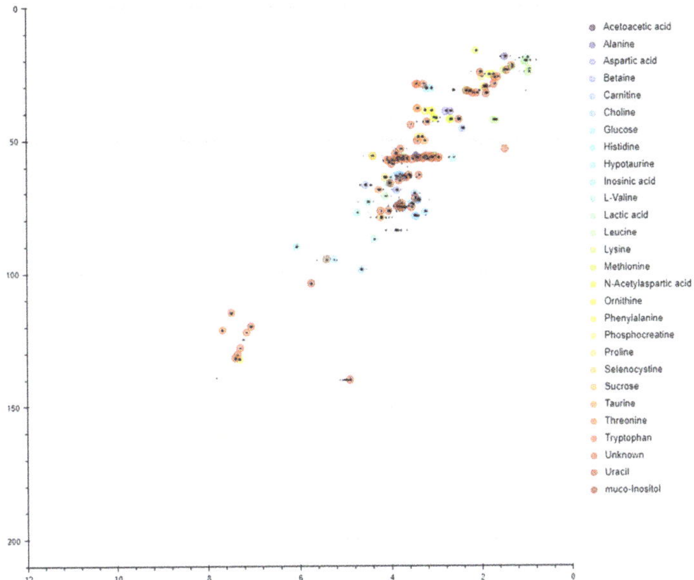

Figure 24. Overlay of experimental HSQC spectra from a metabolite mixture (black correlations) and the outcomes predicted by SMART-Miner (colored correlations). Figure reprinted with permission from Ref. [126]. Copyright 2021 Wiley Periodicals, Inc.

Brougham et al. [127], by employing ANNs on ^{1}H NMR spectra, performed a successful classification of four lung carcinoma cell lines, showing different drug-resistance patterns. The authors chose human lung carcinoma and adenocarcinoma cell lines together with specific drug-resistant daughter lines (Figure 25). The ANN architecture was constructed at first using three layers and the corresponding weights were determined by minimizing the root mean square error. Then, the authors analyzed networks with four layers, two of which are hidden. Their results show that the four-layer structure with two hidden layers provided a 100% successful classification [127]. These data are very interesting in terms of the robustness of the used approach: the cell lines were correctly classified, even though the effects were provoked by the operator and independently from the spectra chosen for training and validation (Figure 25).

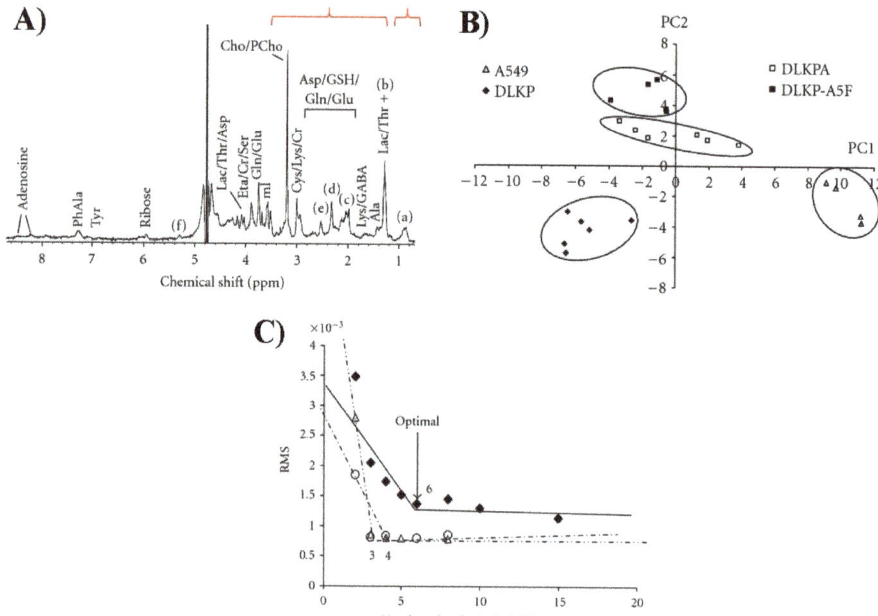

Figure 25. (**A**) Example of ^{1}H NMR spectrum for DLKP lung carcinoma cells. Labeled peak corresponds to (a) CH_3, (b) CH_2, (c) $CH_2CH{=}CH$, (d) CH_2COO, (e) $=CHCH_2CH=$, and (f) HC=CH/CHOCOR. The highlighted intervals at 0.60–1.04 and 1.24–3.56 ppm were used for statistical analysis. (**B**) PCA score plot including data from all four cell lines. (**C**) Residual mean squares error vs. nodes number in the hidden layers, for the 3-layers (full symbols), and in the second (empty triangles) and third (empty circles) layer for the 4-layers networks. Figure reprinted from Ref. [127] under the terms of the Creative Commons Attribution License.

Very recently, Di Donato et al. [128] analyzed serum samples from 94 elderly patients with early stage colorectal cancer and 75 elderly patients with metastatic colorectal cancer. With the aim to separately observe each different molecular components, these authors acquired one-dimensional proton NMR spectra by using three different pulse sequences for each sample: (i) a nuclear Overhauser effect spectroscopy pulse sequence to observe molecules with both low and high molecular weight; (ii) a common spin echo mono-dimensional pulse sequence [129] to observe only lighter metabolites and (iii) a common diffusion-edited pulse sequence to observe only macromolecules [128]. Their results, taking advantage of Kaplan–Meier curves for prognosis and of a PCA-based kNN analysis, allowed distinguishing relapse-free and metastatic cancer groups, with the advantage of obtaining information about the risks in the early stage of the colorectal cancer disease.

Peng et al. [130], by using two-dimensional NMR correlational spectroscopy on the longitudinal (T_1) and transversal components (T_2) of the magnetization relaxation time during its equilibrium recovery, were able to perform a molecular phenotyping of blood with the employment of supervised learning models, including neural networks. In detail, by means of a fast two-dimensional Laplace inversion [117], they obtained T_1–T_2 correlation spectra on a single drop of blood (<5 µL) in a few minutes (Figure 26) with a benchtop-sized NMR spectrometer. Then, they converted the NMR correlational maps for deep image analysis, achieving useful insights for medical decision making by the application of machine learning techniques. In particular, after an initial dimensionality reduction by unsupervised analysis, supervised neural network models were applied to train and predict the data that, at the end, were compared with the diagnostic prediction made by humans. The results showed that ML approaches outperformed the human being and took a much shorter time. Therefore, the authors demonstrated the clinical efficacy of this technique by analyzing human blood in different physiological and pathological conditions, such as oxidation states [130]. Concerning the analysis of different physiological conditions, Figure 26 reports the T_1–T_2 correlational maps of blood cells at oxygenated (a), oxidized (b), and deoxygenated (c) states. Three peaks with different relaxation times values were observed and assigned to the different microenvironments that water experiences in the considered samples of red blood cells. For the obtained maps, the coordinate for the bulk water peak (slowest component) is shown at the upper left of the map indicating T_2 and T_1 relaxations (in ms) and T_1/T_2-ratio, respectively. Instead, the coordinates of the fastest components, due to hydration and bound water molecules [131], are reported close to the corresponding correlation peak (Figure 26).

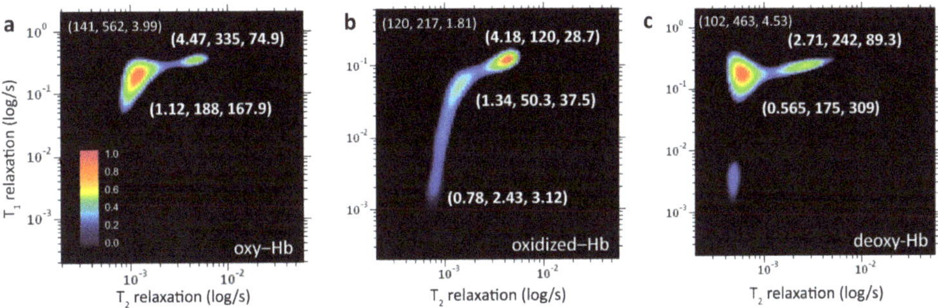

Figure 26. T_1–T_2 correlational maps in false colors of red blood cells at different conditions: oxygenated (**a**), oxidized (**b**), and deoxygenated (**c**). Figure reprinted from Ref. [130] under the terms of the Creative Commons Attribution 4.0 International License.

5. Conclusions and Future Perspective

The role played by each metabolite (in terms of identification and quantification) is essential to validate NMR spectroscopy potentiality in this field. Overall, NMR-based metabolomics coupled with machine learning and neural networks improves its power, especially in the food and biomedical fields, paving the way for innovative and hybrid approaches for deep insights into the metabolic fingerprinting of complex biological matrices. In fact, the number of identified metabolites is very low, and in some cases, the metabolites profile analysis is difficult for the high noise level and the multicolinearity with respect to the genomics case. However, the coupling of genomics and metabolomics tools is still a goal to be achieved. To this purpose, the deep learning and neural network approaches are the best methods to use, although the first step may involve the use of linear discriminant analysis to select a subset of metabolites to be used as input for the neural network analysis in view of an accurate classification as well as the generalizability of the method. Therefore, some efforts are still necessary for applying deep learning approaches on NMR metabolomics data, strictly related to the specific properties of the selected/investigated metabolites, evaluating the dimensionality reduction problems and improving the reliability of the evaluation models.

Author Contributions: Conceptualization, C.C., F.N. and E.F.; methodology, S.V., A.M.M. and G.N.; writing—review and editing, all authors. All authors have read and agreed to the published version of the manuscript.

Funding: This research received no external funding.

Institutional Review Board Statement: Not applicable.

Informed Consent Statement: Not applicable.

Data Availability Statement: No new data were created or analyzed in this study, so data sharing is not applicable to this article.

Conflicts of Interest: The authors declare no conflict of interest.

Abbreviations

The following abbreviations are used in this manuscript:

NMR	Nuclear Magnetic Resonance
MS	Mass Spectrometry
AI	Artificial Intelligence
ML	Machine Learning
DL	Deep Learning
NN	Neural Network
ANN	Artificial Neural Network
DNN	Deep Neural Network
PCA	Principal Component Analysis
PLS	Partial Least Squares
ORA	Over Representation Analysis
FCS	Functional Class Scoring

Appendix A. Technical Aspects

Nuclear magnetic resonance (NMR) is one of the most employed experimental techniques for investigating the wide composition and structural complexity of biological samples. The NMR technique is characterized by high reproducibility and ease of sample preparation and measurement proceedings. NMR is a non-destructive technique able to perform different measurements on the same sample, providing increasingly accurate and detailed information. NMR also allows to reach a quantitative analysis and to carry out in vivo metabolomics studies. Unfortunately, it has a relatively low sensitivity (μM), but, in combination with chromatography, it shows a great potentiality for targeted analysis. However, it is a relatively young experimental technique with continuous development from both the hardware and software point of view (see Ref. [3] for a more details). For instance, cryoprobes [132–134] and magic angle techniques [17,135,136] are today commonly used for improving the signal-to-noise ratio, while AI methods are used both for signal pre-processing, such as baseline optimization [137–139], and for data analysis, as discussed in the main text of this review.

Briefly, the NMR working principle is based on the resonant excitation of the precession dynamics of the nuclear magnetic moment under the effect of a static magnetic field. Nuclei characterized by an odd number of protons and/or neutrons show a magnetic moment, associated to the nuclear spin characterized by the corresponding quantum number (I). Nuclei with $I \neq 0$ possess an intrinsic nuclear magnetic moment (μ) so producing a slight local magnetic field (B_0). Once immersed in an external magnetic field (B), these nuclei, previously randomly orientated, align themselves either in the same or opposite direction of B. These nuclei, subject to B, move in a precessional motion at a frequency called Larmor frequency, which takes on values in the range of 50–900 MHz (see Figure A1). Indeed, it is characteristic for each nucleus and increases with the strength of the external magnetic field B. In this condition, if the system is irradiated with an electromagnetic radiation at the corresponding Larmor frequency (resonance condition), nuclei can absorb the radiation energy, and the nuclear spins can be promoted to a different Zeeman level.

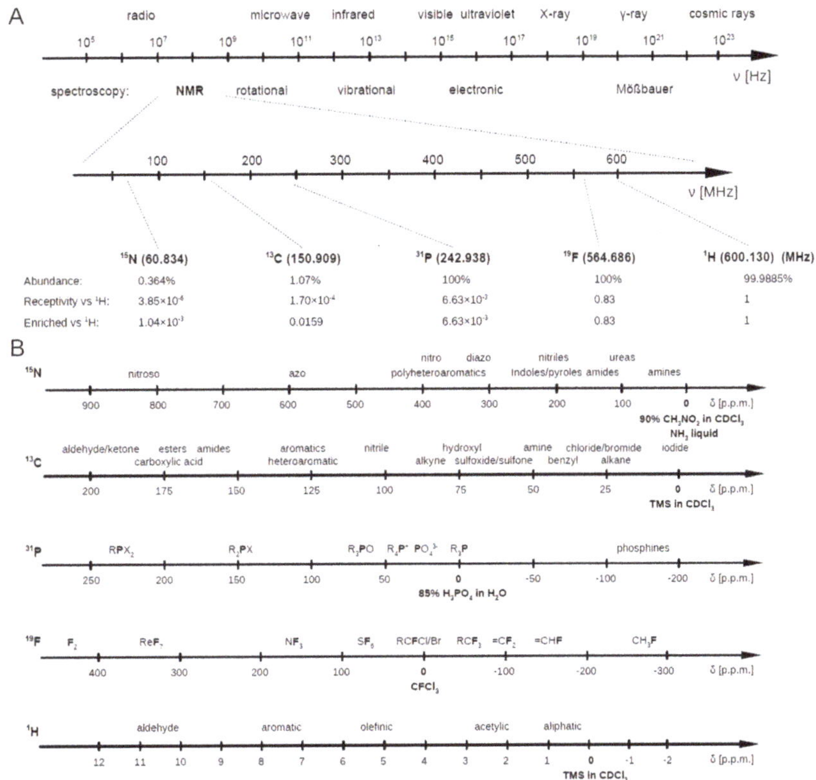

Figure A1. (**A**) Spectroscopies and corresponding frequency ranges. Larmor frequency of most used nuclei for metabolomics analyses with respect to that of the proton when at 600 MHz. (**B**) Parts per million intervals for all these nuclei (^{15}N, ^{13}C, ^{31}P, ^{19}F and ^{1}H) at characteristic chemical environments. Figure reprinted from Ref. [3] under the terms of the Creative Common CC-BY license.

When spins relax toward the fundamental state, they emit a radio frequency (damped in time, called free induction decay (FID)) that well characterizes each nucleus of the system, depending on the corresponding chemical environment that essentially exerts a local magnetic field, causing a shift (chemical shift) from the pure Larmor frequency value. This is commonly indicated by δ and measured in parts per million since the recorded frequency is divided by the spectrometer working frequency such that the spectra acquired with different instruments can be compared. Note that nuclei with I = 0, such as ^{12}C and ^{16}O, are NMR inactive [140,141].

Figure A1B reports the most common NMR active nuclei. Among them, ^{13}C and ^{15}N show a wide chemical shift range, together with a sharp line signal, but their poor natural abundance and the low sensitivity (compared to other nuclei as ^{1}H or ^{19}F) limit their employment in the metabolomic investigation. ^{31}P has a good sensitivity (6.6×10^{-2} relative to ^{1}H) and a wide spectral range, but only few metabolites, such as nucleoside or phospholipids, contain it, restricting its employment to a few compounds. The same comments can be done about ^{19}F.

The high abundance in nature, high sensitivity and relevant gyromagnetic ratio of ^{1}H makes 1D ^{1}H NMR spectra especially useful in the metabolomic investigation. The 1D ^{1}H NMR spectra are fast to record (few minutes) and just the information contained in only one spectrum can provide useful data to identify and quantify from 50 to 100 metabolites [142,143]. In this case, if nuclear spins are totally relaxed and no polarization transfer sequences are applied, the intensity of each acquired proton signal is correlated with the corresponding concentration levels in the molecules, and the area under each peak is directly proportional to the number of ^{1}H constituting the corresponding residue, giving a real distribution of the individual metabolites in the sample mixture. This quantification is possible without previous calibration, thanks to the large linear dynamic range and signal response that characterize proton NMR spectroscopy.

Another important aspect in the analysis of the ^{1}H NMR spectra is the solvent suppression, and in this way, several protocols can be used. Commonly, the protonated solvent can be replaced with a deuterated one; this procedure can also require the lyophilization of the sample and the subsequent dispersion in the deuterated solvent. When this is not possible, the solvent peak can be suppressed by using proper pulse sequences [144]. Regarding the identification of metabolites constituting a biological matrix, when they have a unique and high reproducible fingerprint at specific conditions (pH, solvent, temperature), the non-target strategy can be adopted [145]. This consists of the employment of multimodal models, which clarify how the NMR fingerprint of each sample and among the groups correlate with each other, providing a static analysis. This strategy is very important to give a first overview about the sample composition; however, it is not sufficient to analyze very complex samples. In the latter case, it is more common to adopt the target strategy, which consists in the comparison of the acquired data with available metabolite databases, such as the Human Metabolome Database, Biological Magnetic Resonance Data Bank, Birmingham Metabolite Library, Bbiorefcode (Bruker Biospin Ltd., Billerica, MA, USA) and Chenomx library (Chenomx Inc., Edmonton, AB, USA) [145].

Figure A2 reports a ^{1}H NMR spectrum acquired from human serum at 700 MHz: 55 different metabolites were identified and labeled in the recorded spectrum [3]. In particular, each proton signal can be attributed to the different components of the biofluid, thanks to the high sensitivity of ^{1}H nuclei, its natural abundance and the remarkably narrow line widths, giving a remarkable spectrum resolution. Note that the high intensity of the lactate peak is due to a conversion of the glucose in lactate during the preparation of the sample. To reach a certain assignment of the detected metabolic peaks 1D ^{1}H NMR is sometimes not sufficient. This is due to the relevant numbers of resonances with an ambiguous assignment, and to a peak overlap of the matrix's components. Thus, bi-dimensional (2D) NMR techniques, which investigate the spin–spin correlation among ^{1}H-^{1}H nuclei or with heteroatoms, such as ^{13}C, ^{15}N, ^{31}P, are adopted. In metabolomic studies, typical 2D NMR techniques are ^{1}H-^{1}H correlated spectroscopy (COSY) and total correlation spectroscopy (TOCSY), ^{1}H-^{13}C heteronuclear single quantum coherence (HSQC) and heteronuclear multiple bond correlation (HMBC). HSQC is a great experiment for metabolites identification, which gives information on the direct connectivity between protons and heteroatoms. In particular, the large chemical shift scale of ^{13}C helps to solve the tough issue of the overlapped signals in the proton spectrum, and the variety of HSQC experiments can provide different sets of information on the investigated sample.

For instance, the potentiality of the HSQC technique was proved in the identification of methyl groups of betaine and trimethylamine-N-oxide (TMAO). The proton resonances of methyl groups in TMAO and betaine organic compounds are both close to $\delta = 3.26$ ppm, and thus, the signals are not distinguishable. Instead, the carbon chemical shift of methyl groups in TMAO is assigned at 62.2 ppm, while that of betaine is at 55.8 ppm. This information can be easily obtained by the ^{1}H-^{13}C HSQC experiment (see Figure A3), giving an unambiguous identification of the two organic compounds [145]. HMBC is an appropriate technique to analyze the correlations using the coupling of protons with heteroatoms, which are separated up to four bonds, providing complementary information to that given by HSQC for the structural characterization of metabolites.

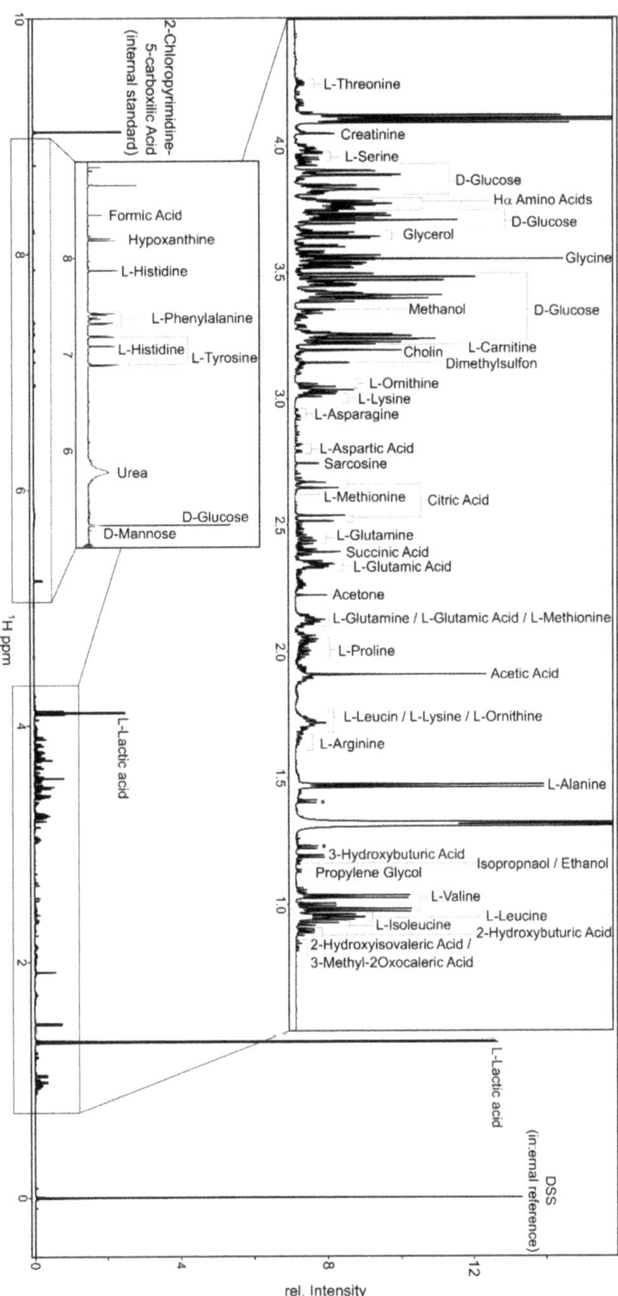

Figure A2. ^{1}H NMR spectrum of ultrafiltered human serum at 700 MHz with the identified compounds labeled above each of the corresponding peaks. Figure reprinted from Ref. [3] under the terms of the Creative Common CC-BY license.

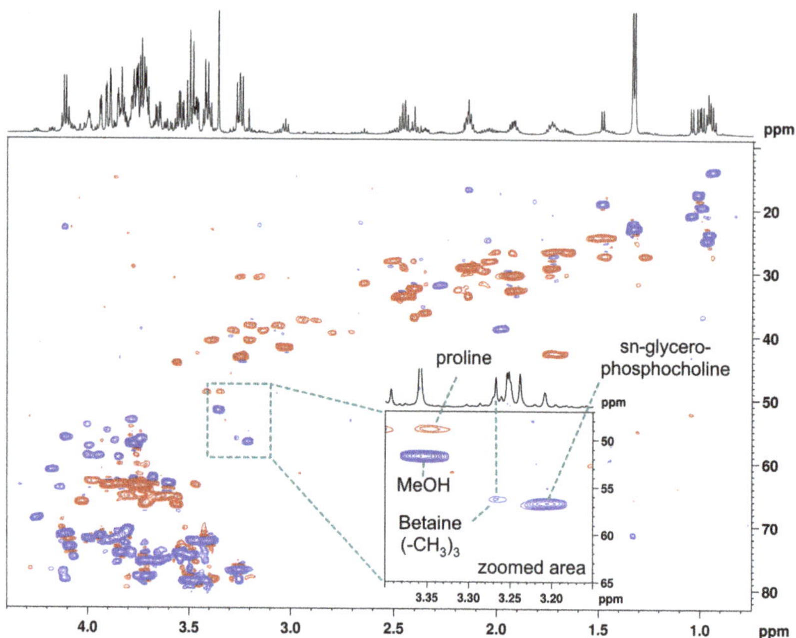

Figure A3. ^{1}H-^{13}C 2D HSQC experiment to identify TMAO and betaine organic compounds of a biological matrix. Figure reprinted from Ref. [145] under the terms of the Creative Common CC-BY license.

Thus, metabolic identification can be easily reached by the combination of 2D-NMR techniques with the metabolite databases. However, in some cases, the concentration of metabolites is very low, and their peaks are often overlapped making their identification difficult, even when employing 2D NMR techniques. In these cases, if the sharp chemical shift of the compounds to identify is known, it is recommended to use a standard (reference) compound, which is added in the concentration range 10–100 µM. For instance, this method was applied to identify the uridine diphosphate (UDP) conjugates, which are present in very low concentrations in cellular extracts with overlapped peaks, but their chemical shift is well known and the signal-to-noise ratio (S/N) has sufficient intensity to be quantified by 1D/2D NMR experiments [145]. Figure A4 shows in details how the spiking of pure compounds into a mixture aids the identification of metabolites within the spectrum and also its quantification by performing peak fitting of the two spectral regions corresponding to UDP-nacetylglucosamin (UDP-Gluc-NAc). Note that the proton signal on the left side (δ = 5.50 ppm) of UDP-Gluc-NAc is overlapped to that of galactose-1-phosphate (Gal-1-P), whereas signals from the uridine group (δ = 5.95 ppm) superimpose with those from UDP-glucose. In addition, without spiking, it is almost impossible to define the shift of the methyl group belonging to UDP-Gluc-Nac acetyl (right region) since there is a big overlap with other signals, such as the multiplets from glutamine and glutamate.

The addition of a standard is also employed to obtain an absolute quantification of the metabolites contained in the sample. Therefore, the estimation of the metabolites concentration can be made by comparing the area of the metabolites NMR peaks with that of the reference sample by the following equation:

$$\frac{M}{S} = \frac{Im}{Is} \times \frac{Ns}{Nm} \tag{A1}$$

in which M and S represent the amounts of the considered metabolite and that of the reference, while Im and Is indicate the area under the curve of corresponding peaks, and Nm and Ns represent the number of protons which contribute to these bands, respectively [145]. To quantify a small set of metabolites whose resonances are well-resolved peaks, also the pulse length-based concentration determination (PULCON) quantitative NMR can be used. It considers that the signal intensity is inversely proportional to the duration of the 90° pulse adopted to excite nuclei [146].

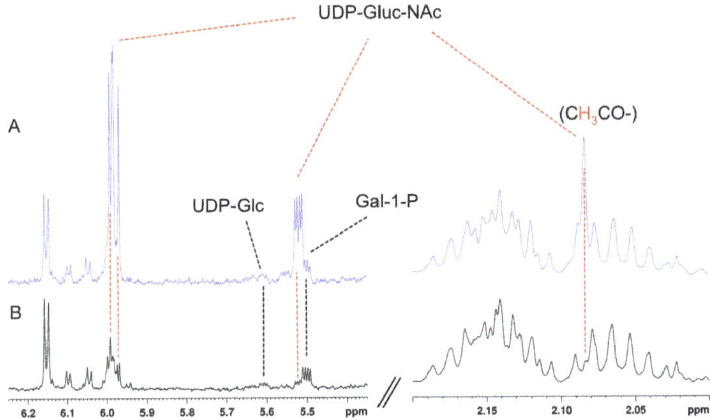

Figure A4. (**A**): The spiking of UDP-nacetylglucosamine (UDP-Gluc-NAc) allows its identification and quantitation. (**B**): The same spectrum of A without the addition of UDP-Gluc-NAc. Figure reprinted from Ref. [145] under the terms of the Creative Common CC-BY license.

Finally, another possibility for quantitative NMR analysis was reached by using a digital standard electronic reference to access in vivo concentration (ERETIC) technique [147]. It consists of the generation of a signal via a second channel of the probe and the addition of it as a pseudo-FID during the acquisition of the proton experiment, resulting in a common NMR signal [148]. Initially, the ERETIC technique required to be calibrated before running the quantification measurements and some hardware rearrangements. Improvements of ERETIC are ERETIC2 (Bruker Biospin, Topspin 3.0) and quantification by artificial signal (QUANTAS) [149].

Considering the complexity of NMR spectrum of metabolites, often, peak integration is not a sufficient method for the quantitative estimation, and in these cases, the deconvolution approach is preferred. It consists in the fit of a target peak of the compound by using the signal acquired from the reference compound [150]. Different specific software for NMR, such as TopSpin (Bruker), MNova (Mestrelab Research), Spectrus Processor (ACD/Labs), Delta (JEOL) and Chenomx NMR Suite (Chenomx Inc.) can be used for this goal. Among them, JEOL Delta is the only one completely free of charge, while Chenomx NMR Suite seems to show the best performance because it is based on a sophisticated targeted profiling technology and on reference libraries containing hundreds of metabolite spectral data, allowing a user-friendly deconvolution of complex NMR spectra [151]. The spectral analysis and deconvolution can also be performed with non-specific software, such as Matlab (The MathWorks, Inc.) or R (The R Foundation).

Several factors (i.e., pulse sequence changes or variation in the repetition time) influence the deconvolution process and its accuracy, including the variety of standard compounds present in the library and the need to repeat the NMR data acquisition in the same experimental conditions. Changes in the pulse sequence and/or the repetition time result in a less accurate fitting. The performance of the deconvolution is also influenced by the protons' bond to nitrogen atoms, also called labile (for instance, the α-protons in amino acids) [145]. These protons fast exchange with the solvent, and this not only makes it difficult to detect them, but also provokes changes in the line shape of the close protons peaks. The result is an attenuated resonance, which does not precisely match with the integral corresponding to the other proton peaks in the considered sample. Another error regards the partial peak saturation of the protons, with a resonance close to the presaturation peak of the solvent (commonly water). This was observed for the anomeric protons of carbohydrates, which are frequently resonant close to the water signal, or also for the CH quartet at δ 4.11 ppm in the lactate spectrum. Beyond these disadvantages, the deconvolution approach is a great and widely employed tool for the metabolomic quantification studies [145].

Generally, successful NMR metabolomics requires statistical analyses, which have become progressively advanced over the years, and are the focus of this review. Dependent and independent parameters are correlated by means of conventional approaches on the basis of the mathematical relationship and, in turn, on model fitting. On the other hand, machine learning approaches group input data based on a cluster classification without any statistical assumption, while deep learning is devoted to find statistical inferences from a large amount of input data. The future of NMR-based metabolomics is to generalize the learning approaches to optimize predictive ability for specific diseases.

References

1. Muthubharathi, B.C.; Gowripriya, T.; Balamurugan, K. Metabolomics: Small molecules that matter more. *Mol. Omics* **2021**, *17*, 210–229. [CrossRef] [PubMed]
2. Zhu, M.; Du, X.; Xu, H.; Yang, S.; Wang, C.; Zhu, Y.; Zhang, T.; Zhao, W. Metabolic profiling of liver and faeces in mice infected with echinococcosis. *Parasites Vectors* **2021**, *14*, 324. [CrossRef] [PubMed]
3. Emwas, A.H.; Roy, R.; McKay, R.T.; Tenori, L.; Saccenti, E.; Gowda, G.A.N.; Raftery, D.; Alahmari, F.; Jaremko, L.; Jaremko, M.; et al. NMR Spectroscopy for Metabolomics Research. *Metabolites* **2019**, *9*, 123. [CrossRef] [PubMed]
4. Onuh, J.O.; Qiu, H. Metabolic Profiling and Metabolites Fingerprints in Human Hypertension: Discovery and Potential. *Metabolites* **2021**, *11*, 687. [CrossRef]
5. Caspani, G.; Sebők, V.; Sultana, N.; Swann, J.R.; Bailey, A. Metabolic phenotyping of opioid and psychostimulant addiction: A novel approach for biomarker discovery and biochemical understanding of the disorder. *Br. J. Pharmacol.* **2021**, 1–29. [CrossRef]
6. Wishart, D.S.; Guo, A.; Oler, E.; Wang, F.; Anjum, A.; Peters, H.; Dizon, R.; Sayeeda, Z.; Tian, S.; Lee, B.L.; et al. HMDB 5.0: The Human Metabolome Database for 2022. *Nucleic Acids Res.* **2022**, *50*, D622–D631. [CrossRef]
7. Ulrich, E.L.; Akutsu, H.; Doreleijers, J.F.; Harano, Y.; Ioannidis, Y.E.; Lin, J.; Livny, M.; Mading, S.; Maziuk, D.; Miller, Z.; et al. BioMagResBank. *Nucleic Acids Res.* **2007**, *36*, D402–D408. [CrossRef]
8. Goodacre, R.; Broadhurst, D.; Smilde, A.K.; Kristal, B.S.; Baker, J.D.; Beger, R.; Bessant, C.; Connor, S.; Capuani, G.; Craig, A.; et al. Proposed minimum reporting standards for data analysis in metabolomics. *Metabolomics* **2007**, *3*, 231–241. [CrossRef]
9. Claridge, T.D. *High-Resolution NMR Techniques in Organic Chemistry*; Elsevier: Amsterdam, The Netherlands, 2016. [CrossRef]
10. Oyedeji, A.B.; Green, E.; Adebiyi, J.A.; Ogundele, O.M.; Gbashi, S.; Adefisoye, M.A.; Oyeyinka, S.A.; Adebo, O.A. Metabolomic approaches for the determination of metabolites from pathogenic microorganisms: A review. *Food Res. Int.* **2021**, *140*, 110042. [CrossRef]
11. Letertre, M.P.M.; Giraudeau, P.; de Tullio, P. Nuclear Magnetic Resonance Spectroscopy in Clinical Metabolomics and Personalized Medicine: Current Challenges and Perspectives. *Front. Mol. Biosci.* **2021**, *8*, 698337. [CrossRef]
12. Emwas, A.H.; Alghrably, M.; Al-Harthi, S.; Poulson, B.G.; Szczepski, K.; Chandra, K.; Jaremko, M. New Advances in Fast Methods of 2D NMR Experiments. In *Nuclear Magnetic Resonance*; IntechOpen: London, UK, 2020. [CrossRef]
13. Deaton, A.; Cartwright, N. Understanding and misunderstanding randomized controlled trials. *Soc. Sci. Med.* **2018**, *210*, 2–21. [CrossRef] [PubMed]
14. Davies, N.M.; Holmes, M.V.; Davey Smith, G. Reading Mendelian randomisation studies: A guide, glossary, and checklist for clinicians. *BMJ* **2018**, *362*, k601. [CrossRef] [PubMed]
15. Teumer, A. Common Methods for Performing Mendelian Randomization. *Front. Cardiovasc. Med.* **2018**, *5*, 51. [CrossRef] [PubMed]
16. Mishra, P.; Biancolillo, A.; Roger, J.M.; Marini, F.; Rutledge, D.N. New data preprocessing trends based on ensemble of multiple preprocessing techniques. *TrAC Trends Anal. Chem.* **2020**, *132*, 116045. [CrossRef]
17. Augustijn, D.; de Groot, H.J.M.; Alia, A. HR-MAS NMR Applications in Plant Metabolomics. *Molecules* **2021**, *26*, 931. [CrossRef]
18. Xu, X.; Xie, Z.; Yang, Z.; Li, D.; Xu, X. A t-SNE Based Classification Approach to Compositional Microbiome Data. *Front. Genet.* **2020**, *11*, 1633. [CrossRef]
19. Worley, B.; Powers, R. Generalized adaptive intelligent binning of multiway data. *Chemom. Intell. Lab. Syst.* **2015**, *146*, 42–46. [CrossRef]

20. Emwas, A.H.; Saccenti, E.; Gao, X.; Mckay, R.; Martins dos Santos, V.; Roy, R.; Wishart, D. Recommended strategies for spectral processing and post-processing of 1D 1H-NMR data of biofluids with a particular focus on urine. *Metabolomics* **2018**, *14*, 31. [CrossRef]
21. Anderson, P.; Reo, N.; Delraso, N.; Doom, T.; Raymer, M. Gaussian binning: A new kernel-based method for processing NMR spectroscopic data for metabolomics. *Metabolomics* **2008**, *4*, 261–272. [CrossRef]
22. Puchades-Carrasco, L.; Palomino-Schätzlein, M.; Pérez-Rambla, C.; Pineda-Lucena, A. Bioinformatics tools for the analysis of NMR metabolomics studies focused on the identification of clinically relevant biomarkers. *Brief. Bioinform.* **2015**, *17*, 541–552. [CrossRef]
23. Hu, J.M.; Sun, H.T. Serum proton NMR metabolomics analysis of human lung cancer following microwave ablation. *Radiat. Oncol.* **2018**, *13*, 40. [CrossRef]
24. Dieterle, F.; Ross, A.; Schlotterbeck, G.; Senn, H. Probabilistic Quotient Normalization as Robust Method to Account for Dilution of Complex Biological Mixtures. Application in 1H NMR Metabonomics. *Anal. Chem.* **2006**, *78*, 4281–4290. [CrossRef] [PubMed]
25. Liu, Z.; Abbas, A.; Jing, B.Y.; Gao, X. WaVPeak: Picking NMR peaks through wavelet-based smoothing and volume-based filtering. *Bioinformatics* **2012**, *28*, 914–920. [CrossRef] [PubMed]
26. MacDonald, R.; Sokolenko, S. Detection of highly overlapping peaks via adaptive apodization. *J. Magn. Reson.* **2021**, *333*, 107104. [CrossRef] [PubMed]
27. Dona, A.C.; Kyriakides, M.; Scott, F.; Shephard, E.A.; Varshavi, D.; Veselkov, K.; Everett, J.R. A guide to the identification of metabolites in NMR-based metabonomics/metabolomics experiments. *Comput. Struct. Biotechnol. J.* **2016**, *14*, 135–153. [CrossRef]
28. Khalili, B.; Tomasoni, M.; Mattei, M.; Mallol Parera, R.; Sonmez, R.; Krefl, D.; Rueedi, R.; Bergmann, S. Automated Analysis of Large-Scale NMR Data Generates Metabolomic Signatures and Links Them to Candidate Metabolites. *J. Proteome Res.* **2019**, *18*, 3360–3368. [CrossRef]
29. Jaadi, Z. A Step-by-Step Explanation of Principal Component Analysis (PCA). Available online: https://builtin.com/data-science/step-step-explanation-principal-component-analysis (accessed on 8 January 2022).
30. AG, S. What Is Principal Component Analysis (PCA) and How It Is Used? Available online: https://www.sartorius.com/en/knowledge/science-snippets/what-is-principal-component-analysis-pca-and-how-it-is-used-507186 (accessed on 8 January 2022).
31. Jolliffe, I.T.; Cadima, J. Principal component analysis: A review and recent developments. *Philos. Trans. R. Soc. A Math. Phys. Eng. Sci.* **2016**, *374*, 20150202. [CrossRef]
32. Parsons, H.M.; Ludwig, C.; Günther, U.L.; Viant, M.R. Improved classification accuracy in 1- and 2-dimensional NMR metabolomics data using the variance stabilising generalised logarithm transformation. *BMC Bioinform.* **2007**, *8*, 234. [CrossRef]
33. Izquierdo-Garcia, J.L.; del Barrio, P.C.; Campos-Olivas, R.; Villar-Hernández, R.; Prat-Aymerich, C.; Souza-Galvão, M.L.D.; Jiménez-Fuentes, M.A.; Ruiz-Manzano, J.; Stojanovic, Z.; González, A.; et al. Discovery and validation of an NMR-based metabolomic profile in urine as TB biomarker. *Sci. Rep.* **2020**, *10*, 22317. [CrossRef]
34. Shiokawa, Y.; Date, Y.; Kikuchi, J. Application of kernel principal component analysis and computational machine learning to exploration of metabolites strongly associated with diet. *Sci. Rep.* **2018**, *8*, 3426. [CrossRef]
35. Halouska, S.; Powers, R. Negative impact of noise on the principal component analysis of NMR data. *J. Magn. Reson.* **2006**, *178*, 88–95. [CrossRef] [PubMed]
36. Rutledge, D.N.; Roger, J.M.; Lesnoff, M. Different Methods for Determining the Dimensionality of Multivariate Models. *Front. Anal. Sci.* **2021**, *1*, 754447. [CrossRef]
37. Smilde, A.K.; Jansen, J.J.; Hoefsloot, H.C.J.; Lamers, R.J.A.N.; van der Greef, J.; Timmerman, M.E. ANOVA-simultaneous component analysis (ASCA): A new tool for analyzing designed metabolomics data. *Bioinformatics* **2005**, *21*, 3043–3048. [CrossRef]
38. Lemanska, A.; Grootveld, M.; Silwood, C.J.L.; Brereton, R.G. Chemometric variance analysis of NMR metabolomics data on the effects of oral rinse on saliva. *Metabolomics* **2012**, *8*, 64–80. [CrossRef]
39. Puig-Castellví, F.; Alfonso, I.; Piña, B.; Tauler, R. 1H NMR metabolomic study of auxotrophic starvation in yeast using Multivariate Curve Resolution-Alternating Least Squares for Pathway Analysis. *Sci. Rep.* **2016**, *6*, 30982. [CrossRef]
40. Trepalin, S.V.; Yarkov, A.V. Hierarchical Clustering of Large Databases and Classification of Antibiotics at High Noise Levels. *Algorithms* **2008**, *1*, 183–200. [CrossRef]
41. Tiwari, P.; Madabhushi, A.; Rosen, M. A Hierarchical Unsupervised Spectral Clustering Scheme for Detection of Prostate Cancer from Magnetic Resonance Spectroscopy (MRS). In *Medical Image Computing and Computer-Assisted Intervention—MICCAI 2007*; Ayache, N., Ourselin, S., Maeder, A., Eds.; Springer: Berlin/Heidelberg, Germany, 2007; pp. 278–286.
42. Čuperlović Culf, M.; Belacel, N.; Culf, A.S.; Chute, I.C.; Ouellette, R.J.; Burton, I.W.; Karakach, T.K.; Walter, J.A. NMR metabolic analysis of samples using fuzzy K-means clustering. *Magn. Reson. Chem.* **2009**, *47*, S96–S104. [CrossRef]
43. Zou, X.; Holmes, E.; Nicholson, J.K.; Loo, R.L. Statistical HOmogeneous Cluster SpectroscopY (SHOCSY): An Optimized Statistical Approach for Clustering of 1H NMR Spectral Data to Reduce Interference and Enhance Robust Biomarkers Selection. *Anal. Chem.* **2014**, *86*, 5308–5315. [CrossRef]
44. Gülseçen, S.; Sharma, S.; Akadal, E. *Who Runs the World: Data*; Istanbul University Press: Istanbul, Turkey, 2020. [CrossRef]
45. Schonlau, M. Visualizing non-hierarchical and hierarchical cluster analyses with clustergrams. *Comput. Stat.* **2004**, *19*, 95–111. [CrossRef]
46. Yim, O.; Ramdeen, K.T. Hierarchical Cluster Analysis: Comparison of Three Linkage Measures and Application to Psychological Data. *Quant. Methods Psychol.* **2015**, *11*, 8–21. [CrossRef]

47. Zhang, Z.; Murtagh, F.; Poucke, S.V.V.; Lin, S.; Lan, P. Hierarchical cluster analysis in clinical research with heterogeneous study population: Highlighting its visualization with R. *Ann. Transl. Med.* **2017**, *5*, 75. [CrossRef] [PubMed]
48. Richard, V.; Conotte, R.; Mayne, D.; Colet, J.M. Does the 1H-NMR plasma metabolome reflect the host-tumor interactions in human breast cancer? *Oncotarget* **2017**, *8*, 49915–49930. [CrossRef] [PubMed]
49. Selvaratnam, R.; Chowdhury, S.; VanSchouwen, B.; Melacini, G. Mapping allostery through the covariance analysis of NMR chemical shifts. *Proc. Natl. Acad. Sci. USA* **2011**, *108*, 6133–6138. [CrossRef] [PubMed]
50. Kohonen, T. *Self-Organizing Maps*, 3rd ed.; Springer: Berlin/Heidelberg, Germany, 2001.
51. Kaski, S. Data exploration using self-organizing maps. In *Acta Polytechnica Scandinavica: Mathematics, Computing and Management in Engineering Series no. 82*; Finnish Academy of Technology: Espoo, Finland, 1997.
52. Zheng, H.; Ji, J.; Zhao, L.; Chen, M.; Shi, A.; Pan, L.; Huang, Y.; Zhang, H.; Dong, B.; Gao, H. Prediction and diagnosis of renal cell carcinoma using nuclear magnetic resonance-based serum metabolomics and self-organizing maps. *Oncotarget* **2016**, *7*, 59189–59198. [CrossRef]
53. Akdemir, D.; Rio, S.; Isidro y Sánchez, J. TrainSel: An R Package for Selection of Training Populations. *Front. Genet.* **2021**, *12*, 607. [CrossRef]
54. Migdadi, L.; Lambert, J.; Telfah, A.; Hergenröder, R.; Wöhler, C. Automated metabolic assignment: Semi-supervised learning in metabolic analysis employing two dimensional Nuclear Magnetic Resonance (NMR). *Comput. Struct. Biotechnol. J.* **2021**, *19*, 5047–5058. [CrossRef]
55. Alonso-Salces, R.M.; Gallo, B.; Collado, M.I.; Sasía-Arriba, A.; Viacava, G.E.; García-González, D.L.; Gallina Toschi, T.; Servili, M.; Ángel Berrueta, L. 1H–NMR fingerprinting and supervised pattern recognition to evaluate the stability of virgin olive oil during storage. *Food Control* **2021**, *123*, 107831. [CrossRef]
56. Suppers, A.; Gool, A.J.v.; Wessels, H.J.C.T. Integrated Chemometrics and Statistics to Drive Successful Proteomics Biomarker Discovery. *Proteomes* **2018**, *6*, 20. [CrossRef]
57. Biswas, S.; Bordoloi, M.; Purkayastha, B. Review on Feature Selection and Classification using Neuro-Fuzzy Approaches. *Int. J. Appl. Evol. Comput.* **2016**, *7*, 28–44. [CrossRef]
58. Smola, A.J.; Schölkopf, B. A tutorial on support vector regression. *Stat. Comput.* **2004**, *14*, 199–222. [CrossRef]
59. Altman, N.S. An Introduction to Kernel and Nearest-Neighbor Nonparametric Regression. *Am. Stat.* **1992**, *46*, 175–185. [CrossRef]
60. Venkatesan, P.; Dharuman, C.; Gunasekaran, S. A Comparative Study of Principal Component Regression and Partial least Squares Regression with Application to FTIR Diabetes Data. *Indian J. Sci. Technol.* **2011**, *4*, 740–746. [CrossRef]
61. Wold, S.; Ruhe, A.; Wold, H.; Dunn, W.J., III. The Collinearity Problem in Linear Regression. The Partial Least Squares (PLS) Approach to Generalized Inverses. *SIAM J. Sci. Stat. Comput.* **1984**, *5*, 735–743. [CrossRef]
62. Lee, L.C.; Liong, C.Y.; Jemain, A.A. Partial least squares-discriminant analysis (PLS-DA) for classification of high-dimensional (HD) data: A review of contemporary practice strategies and knowledge gaps. *Analyst* **2018**, *143*, 3526–3539. [CrossRef]
63. Song, W.; Wang, H.; Maguire, P.; Nibouche, O. Nearest clusters based partial least squares discriminant analysis for the classification of spectral data. *Anal. Chim. Acta* **2018**, *1009*, 27–38. [CrossRef]
64. Traquete, F.; Luz, J.; Cordeiro, C.; Sousa Silva, M.; Ferreira, A.E.N. Binary Simplification as an Effective Tool in Metabolomics Data Analysis. *Metabolites* **2021**, *11*, 788. [CrossRef]
65. Jiménez-Carvelo, A.M.; Martín-Torres, S.; Ortega-Gavilán, F.; Camacho, J. PLS-DA vs sparse PLS-DA in food traceability. A case study: Authentication of avocado samples. *Talanta* **2021**, *224*, 121904. [CrossRef]
66. Gabrielsson, J.; Jonsson, H.; Airiau, C.; Schmidt, B.; Escott, R.; Trygg, J. OPLS methodology for analysis of pre-processing effects on spectroscopic data. *Chemom. Intell. Lab. Syst.* **2006**, *84*, 153–158. [CrossRef]
67. Embade, N.; Cannet, C.; Diercks, T.; Gil-Redondo, R.; Bruzzone, C.; Ansó, S.; Echevarría, L.R.; Ayucar, M.M.M.; Collazos, L.; Lodoso, B.; et al. NMR-based newborn urine screening for optimized detection of inherited errors of metabolism. *Sci. Rep.* **2019**, *9*, 13067. [CrossRef]
68. Huang, S.; Cai, N.; Pacheco, P.P.; Narrandes, S.; Wang, Y.; Xu, W. Applications of Support Vector Machine (SVM) Learning in Cancer Genomics. *Cancer Genom. Proteom.* **2018**, *15*, 41–51. [CrossRef]
69. Ghosh, T.; Zhang, W.; Ghosh, D.; Kechris, K. Predictive Modeling for Metabolomics Data. In *Computational Methods and Data Analysis for Metabolomics*; Li, S., Ed.; Springer: New York, NY, USA, 2020; pp. 313–336. [CrossRef]
70. Zhang, T.; Chen, C.; Xie, K.; Wang, J.; Pan, Z. Current State of Metabolomics Research in Meat Quality Analysis and Authentication. *Foods* **2021**, *10*, 2388. [CrossRef] [PubMed]
71. Broadhurst, D.I.; Kell, D.B. Statistical strategies for avoiding false discoveries in metabolomics and related experiments. *Metabolomics* **2006**, *2*, 171–196. [CrossRef]
72. Westerhuis, J.A.; Hoefsloot, H.C.J.; Smit, S.; Vis, D.J.; Smilde, A.K.; van Velzen, E.J.J.; van Duijnhoven, J.P.M.; van Dorsten, F.A. Assessment of PLSDA cross validation. *Metabolomics* **2008**, *4*, 81–89. [CrossRef]
73. Wehrens, R.; Putter, H.; Buydens, L.M. The bootstrap: A tutorial. *Chemom. Intell. Lab. Syst.* **2000**, *54*, 35–52. [CrossRef]
74. Wieder, C.; Frainay, C.; Poupin, N.; Rodríguez-Mier, P.; Vinson, F.; Cooke, J.; Lai, R.P.; Bundy, J.G.; Jourdan, F.; Ebbels, T. Pathway analysis in metabolomics: Recommendations for the use of over-representation analysis. *PLoS Comput. Biol.* **2021**, *17*, e1009105. [CrossRef]
75. Khatri, P.; Sirota, M.; Butte, A.J. Ten Years of Pathway Analysis: Current Approaches and Outstanding Challenges. *PLoS Comput. Biol.* **2012**, *8*, e1002375. [CrossRef]

76. Marco-Ramell, A.; Palau, M.; Alay, A.; Tulipani, S.; Urpi-Sarda, M.; Sánchez-Pla, A.; Andres-Lacueva, C. Evaluation and comparison of bioinformatic tools for the enrichment analysis of metabolomics data. *BMC Bioinform.* **2018**, *19*, 1. [CrossRef]
77. Karnovsky, A.; Li, S. Pathway Analysis for Targeted and Untargeted Metabolomics. *Methods Mol. Biol.* **2020**, *2104*, 387–400.
78. Nguyen, T.M.; Shafi, A.; Nguyen, T.; Draghici, S. Identifying significantly impacted pathways: A comprehensive review and assessment. *Genome Biol.* **2019**, *20*, 203. [CrossRef]
79. García-Campos, M.A.; Espinal-Enríquez, J.; Hernández-Lemus, E. Pathway Analysis: State of the Art. *Front. Physiol.* **2015**, *6*, 383. [CrossRef]
80. Liu, Y.; Xu, X.; Deng, L.; Cheng, K.K.; Xu, J.; Raftery, D.; Dong, J. A Novel Network Modelling for Metabolite Set Analysis: A Case Study on CRC Metabolomics. *IEEE Access* **2020**, *8*, 106425–106436. [CrossRef]
81. Mitrea, C.; Taghavi, Z.; Bokanizad, B.; Hanoudi, S.; Tagett, R.; Donato, M.; Voichita, C.; Draghici, S. Methods and approaches in the topology-based analysis of biological pathways. *Front. Physiol.* **2013**, *4*, 278. [CrossRef] [PubMed]
82. Ihnatova, I.; Popovici, V.; Budinska, E. A critical comparison of topology-based pathway analysis methods. *PLoS ONE* **2018**, *13*, e0191154. [CrossRef] [PubMed]
83. Ma, J.; Shojaie, A.; Michailidis, G. A comparative study of topology-based pathway enrichment analysis methods. *BMC Bioinform.* **2019**, *20*, 546. [CrossRef]
84. Chagoyen, M.; Pazos, F. Tools for the functional interpretation of metabolomic experiments. *Brief. Bioinform.* **2012**, *14*, 737–744. [CrossRef]
85. Huang, D.W.; Sherman, B.T.; Lempicki, R.A. Bioinformatics enrichment tools: Paths toward the comprehensive functional analysis of large gene lists. *Nucleic Acids Res.* **2008**, *37*, 1–13. [CrossRef]
86. Emwas, A.H.M. The Strengths and Weaknesses of NMR Spectroscopy and Mass Spectrometry with Particular Focus on Metabolomics Research. In *Methods in Molecular Biology*; Springer: New York, NY, USA, 2015; pp. 161–193. [CrossRef]
87. Pavlidis, P.; Qin, J.; Arango, V.; Mann, J.J.; Sibille, E. Using the Gene Ontology for Microarray Data Mining: A Comparison of Methods and Application to Age Effects in Human Prefrontal Cortex. *Neurochem. Res.* **2004**, *29*, 1213–1222. [CrossRef]
88. Al-Shahrour, F.; Díaz-Uriarte, R.; Dopazo, J. Discovering molecular functions significantly related to phenotypes by combining gene expression data and biological information. *Bioinformatics* **2005**, *21*, 2988–2993. [CrossRef]
89. Goeman, J.J.; van de Geer, S.A.; de Kort, F.; van Houwelingen, H.C. A global test for groups of genes: Testing association with a clinical outcome. *Bioinformatics* **2004**, *20*, 93–99. [CrossRef]
90. Subramanian, A.; Tamayo, P.; Mootha, V.K.; Mukherjee, S.; Ebert, B.L.; Gillette, M.A.; Paulovich, A.; Pomeroy, S.L.; Golub, T.R.; Lander, E.S.; et al. Gene set enrichment analysis: A knowledge-based approach for interpreting genome-wide expression profiles. *Proc. Natl. Acad. Sci. USA* **2005**, *102*, 15545–15550. [CrossRef]
91. Tian, L.; Greenberg, S.A.; Kong, S.W.; Altschuler, J.; Kohane, I.S.; Park, P.J. Discovering statistically significant pathways in expression profiling studies. *Proc. Natl. Acad. Sci. USA* **2005**, *102*, 13544–13549. [CrossRef] [PubMed]
92. Kim, S.Y.; Volsky, D.J. PAGE: Parametric Analysis of Gene Set Enrichment. *BMC Bioinform.* **2005**, *6*, 144. 1471-2105-6-144. [CrossRef] [PubMed]
93. Jiang, Z.; Gentleman, R. Extensions to gene set enrichment. *Bioinformatics* **2006**, *23*, 306–313. [CrossRef] [PubMed]
94. Kong, S.W.; Pu, W.T.; Park, P.J. A multivariate approach for integrating genome-wide expression data and biological knowledge. *Bioinformatics* **2006**, *22*, 2373–2380. [CrossRef]
95. Barry, W.T.; Nobel, A.B.; Wright, F.A. Significance analysis of functional categories in gene expression studies: A structured permutation approach. *Bioinformatics* **2005**, *21*, 1943–1949. [CrossRef]
96. Efron, B.; Tibshirani, R. On testing the significance of sets of genes. *Ann. Appl. Stat.* **2007**, *1*, 107–129. [CrossRef]
97. Glazko, G.V.; Emmert-Streib, F. Unite and conquer: Univariate and multivariate approaches for finding differentially expressed gene sets. *Bioinformatics* **2009**, *25*, 2348–2354. [CrossRef]
98. Koza, J.R.; Mydlowec, W.; Lanza, G.; Yu, J.; Keane, M.A. Reverse Engineering of Metabolic Pathways From Observed Data Using Genetic Programming. *Pac. Symp. Biocomput.* **2001**, 434–445. [CrossRef]
99. Schmidt, M.D.; Vallabhajosyula, R.R.; Jenkins, J.W.; Hood, J.E.; Soni, A.S.; Wikswo, J.P.; Lipson, H. Automated refinement and inference of analytical models for metabolic networks. *Phys. Biol.* **2011**, *8*, 055011. [CrossRef]
100. Qi, Q.; Li, J.; Cheng, J. Reconstruction of metabolic pathways by combining probabilistic graphical model-based and knowledge-based methods. *BMC Proc.* **2014**, *8*, S5. [CrossRef]
101. Xia, J.; Wishart, D.S. Web-based inference of biological patterns, functions and pathways from metabolomic data using MetaboAnalyst. *Nat. Protoc.* **2011**, *6*, 743–760. [CrossRef] [PubMed]
102. Damiani, C.; Gaglio, D.; Sacco, E.; Alberghina, L.; Vanoni, M. Systems metabolomics: From metabolomic snapshots to design principles. *Curr. Opin. Biotechnol.* **2020**, *63*, 190–199. [CrossRef]
103. Kim, H.I.; Han, K.Y. Urban Flood Prediction Using Deep Neural Network with Data Augmentation. *Water* **2020**, *12*, 899. [CrossRef]
104. Sarker, I.H. Deep Learning: A Comprehensive Overview on Techniques, Taxonomy, Applications and Research Directions. *SN Comput. Sci.* **2021**, *2*, 420. [CrossRef] [PubMed]
105. François-Lavet, V.; Henderson, P.; Islam, R.; Bellemare, M.G.; Pineau, J. An Introduction to Deep Reinforcement Learning. *Found. Trends® Mach. Learn.* **2018**, *11*, 219–354. [CrossRef]

106. Le, T.L.; Huynh, T.T.; Hong, S.K.; Lin, C.M. Hybrid Neural Network Cerebellar Model Articulation Controller Design for Non-linear Dynamic Time-Varying Plants. *Front. Neurosci.* **2020**, *14*, 695. [CrossRef] [PubMed]
107. Arabasadi, Z.; Alizadehsani, R.; Roshanzamir, M.; Moosaei, H.; Yarifard, A.A. Computer aided decision making for heart disease detection using hybrid neural network-Genetic algorithm. *Comput. Methods Programs Biomed.* **2017**, *141*, 19–26. [CrossRef]
108. Fan, X.; Wang, X.; Jiang, M.; Pei, Z.; Qiao, S. An Improved Stacked Autoencoder for Metabolomic Data Classification. *Comput. Intell. Neurosci.* **2021**, *2021*, 1051172. [CrossRef]
109. Zhu, L.; Spachos, P.; Pensini, E.; Plataniotis, K.N. Deep learning and machine vision for food processing: A survey. *Curr. Res. Food Sci.* **2021**, *4*, 233–249. [CrossRef]
110. Sakib, S.; Ahmed, N.; Kabir, A.J.; Ahmed, H. An Overview of Convolutional Neural Network: Its Architecture and Applications. *Preprints* **2018**, 2018110546. [CrossRef]
111. Gil-Solsona, R.; Álvarez-Muñoz, D.; Serra-Compte, A.; Rodríguez-Mozaz, S. (Xeno)metabolomics for the evaluation of aquatic organism's exposure to field contaminated water. *Trends Environ. Anal. Chem.* **2021**, *31*, e00132. [CrossRef]
112. Yang, B.; Zhang, C.; Cheng, S.; Li, G.; Griebel, J.; Neuhaus, J. Novel Metabolic Signatures of Prostate Cancer Revealed by 1H-NMR Metabolomics of Urine. *Diagnostics* **2021**, *11*, 149. [CrossRef] [PubMed]
113. Mandrone, M.; Chiocchio, I.; Barbanti, L.; Tomasi, P.; Tacchini, M.; Poli, F. Metabolomic Study of Sorghum (Sorghum bicolor) to Interpret Plant Behavior under Variable Field Conditions in View of Smart Agriculture Applications. *J. Agric. Food Chem.* **2021**, *69*, 1132–1145. [CrossRef]
114. Nunes, C.A.; Alvarenga, V.O.; de Souza Sant'Ana, A.; Santos, J.S.; Granato, D. The use of statistical software in food science and technology: Advantages, limitations and misuses. *Food Res. Int.* **2015**, *75*, 270–280. [CrossRef] [PubMed]
115. Class, L.C.; Kuhnen, G.; Rohn, S.; Kuballa, J. Diving Deep into the Data: A Review of Deep Learning Approaches and Potential Applications in Foodomics. *Foods* **2021**, *10*, 1803. [CrossRef] [PubMed]
116. Greer, M.; Chen, C.; Mandal, S. Automated classification of food products using 2D low-field NMR. *J. Magn. Reson.* **2018**, *294*, 44–58. [CrossRef] [PubMed]
117. Song, Y.Q.; Venkataramanan, L.; Hürlimann, M.; Flaum, M.; Frulla, P.; Straley, C. T1–T2 Correlation Spectra Obtained Using a Fast Two-Dimensional Laplace Inversion. *J. Magn. Reson.* **2002**, *154*, 261–268. [CrossRef]
118. Date, Y.; Kikuchi, J. Application of a Deep Neural Network to Metabolomics Studies and Its Performance in Determining Important Variables. *Anal. Chem.* **2018**, *90*, 1805–1810. [CrossRef]
119. Wang, D.; Greenwood, P.; Klein, M.S. Deep Learning for Rapid Identification of Microbes Using Metabolomics Profiles. *Metabolites* **2021**, *11*, 863. [CrossRef]
120. Ebrahimnejad, H.; Ebrahimnejad, H.; Salajegheh, A.; Barghi, H. Use of Magnetic Resonance Imaging in Food Quality Control: A Review. *J. Biomed. Phys. Eng.* **2018**, *8*, 127–132. [CrossRef]
121. Caballero, D.; Pérez-Palacios, T.; Caro, A.; Amigo, J.M.; Dahl, A.B.; ErsbØll, B.K.; Antequera, T. Prediction of pork quality parameters by applying fractals and data mining on MRI. *Food Res. Int.* **2017**, *99*, 739–747. [CrossRef] [PubMed]
122. Teimouri, N.; Omid, M.; Mollazade, K.; Mousazadeh, H.; Alimardani, R.; Karstoft, H. On-line separation and sorting of chicken portions using a robust vision-based intelligent modelling approach. *Biosyst. Eng.* **2018**, *167*, 8–20. [CrossRef]
123. Ribeiro, F.D.S.; Caliva, F.; Swainson, M.; Gudmundsson, K.; Leontidis, G.; Kollias, S. An adaptable deep learning system for optical character verification in retail food packaging. In Proceedings of the 2018 IEEE Conference on Evolving and Adaptive Intelligent Systems (EAIS), Kallithea Rhodes, Greece, 25–27 May 2018. [CrossRef]
124. Grapov, D.; Fahrmann, J.; Wanichthanarak, K.; Khoomrung, S. Rise of Deep Learning for Genomic, Proteomic, and Metabolomic Data Integration in Precision Medicine. *OMICS J. Integr. Biol.* **2018**, *22*, 630–636. [CrossRef]
125. Cao, C.; Liu, F.; Tan, H.; Song, D.; Shu, W.; Li, W.; Zhou, Y.; Bo, X.; Xie, Z. Deep Learning and Its Applications in Biomedicine. *Genom. Proteom. Bioinform.* **2018**, *16*, 17–32. [CrossRef] [PubMed]
126. Kim, H.W.; Zhang, C.; Cottrell, G.W.; Gerwick, W.H. SMART-Miner: A convolutional neural network-based metabolite identification from ^{1}H-^{13}C HSQC spectra. *Magn. Reson. Chem.* **2021**. [CrossRef] [PubMed]
127. Brougham, D.F.; Ivanova, G.; Gottschalk, M.; Collins, D.M.; Eustace, A.J.; O'Connor, R.; Havel, J. Artificial Neural Networks for Classification in Metabolomic Studies of Whole Cells Using1H Nuclear Magnetic Resonance. *J. Biomed. Biotechnol.* **2011**, *2011*, 158094. [CrossRef]
128. Di Donato, S.; Vignoli, A.; Biagioni, C.; Malorni, L.; Mori, E.; Tenori, L.; Calamai, V.; Parnofiello, A.; Di Pierro, G.; Migliaccio, I.; et al. A Serum Metabolomics Classifier Derived from Elderly Patients with Metastatic Colorectal Cancer Predicts Relapse in the Adjuvant Setting. *Cancers* **2021**, *13*, 2762. [CrossRef]
129. *Encyclopedia of Spectroscopy and Spectrometry*; Elsevier: Amsterdam, The Netherlands, 2017. [CrossRef]
130. Peng, W.K.; Ng, T.T.; Loh, T.P. Machine learning assistive rapid, label-free molecular phenotyping of blood with two-dimensional NMR correlational spectroscopy. *Commun. Biol.* **2020**, *3*, 535. [CrossRef]
131. Corsaro, C.; Mallamace, D.; Neri, G.; Fazio, E. Hydrophilicity and hydrophobicity: Key aspects for biomedical and technological purposes. *Phys. A Stat. Mech. Its Appl.* **2021**, *580*, 126189. [CrossRef]
132. Chandra, K.; Al-Harthi, S.; Sukumaran, S.; Almulhim, F.; Emwas, A.H.; Atreya, H.S.; Jaremko, Ł.; Jaremko, M. NMR-based metabolomics with enhanced sensitivity. *RSC Adv.* **2021**, *11*, 8694–8700. [CrossRef]
133. Crook, A.A.; Powers, R. Quantitative NMR-Based Biomedical Metabolomics: Current Status and Applications. *Molecules* **2020**, *25*, 5128. [CrossRef] [PubMed]

134. Salmerón, A.M.; Tristán, A.I.; Abreu, A.C.; Fernández, I. Serum Colorectal Cancer Biomarkers Unraveled by NMR Metabolomics: Past, Present, and Future. *Anal. Chem.* **2021**, *94*, 417–430. [CrossRef] [PubMed]
135. Corsaro, C.; Cicero, N.; Mallamace, D.; Vasi, S.; Naccari, C.; Salvo, A.; Giofrè, S.V.; Dugo, G. HR-MAS and NMR towards Foodomics. *Food Res. Int.* **2016**, *89*, 1085–1094. [CrossRef]
136. Corsaro, C.; Fazio, E.; Mallamace, D. Direct Analysis in Foodomics: NMR approaches. In *Comprehensive Foodomics*; Elsevier: Amsterdam, The Netherlands, 2021; pp. 517–535. [CrossRef]
137. Chen, D.; Wang, Z.; Guo, D.; Orekhov, V.; Qu, X. Review and Prospect: Deep Learning in Nuclear Magnetic Resonance Spectroscopy. *Chem.—A Eur. J.* **2020**, *26*, 10391–10401. [CrossRef]
138. Cobas, C. NMR signal processing, prediction, and structure verification with machine learning techniques. *Magn. Reson. Chem.* **2020**, *58*, 512–519. [CrossRef]
139. Helin, R.; Indahl, U.G.; Tomic, O.; Liland, K.H. On the possible benefits of deep learning for spectral preprocessing. *J. Chemom.* **2022**, *26*, e3374. [CrossRef]
140. Silverstein, R.M.; Webster, F.X.; Kiemle, D.J.; Bryce, D.L. *Spectrometric Identification of Organic Compounds*, 8th ed.; Wiley: Hoboken, NJ, USA, 2014.
141. Bisht, B.; Kumar, V.; Gururani, P.; Tomar, M.S.; Nanda, M.; Vlaskin, M.S.; Kumar, S.; Kurbatova, A. The potential of nuclear magnetic resonance (NMR) in metabolomics and lipidomics of microalgae- a review. *Arch. Biochem. Biophys.* **2021**, *710*, 108987. [CrossRef]
142. Holmes, E.; Nicholls, A.W.; Lindon, J.C.; Connor, S.C.; Connelly, J.C.; Haselden, J.N.; Damment, S.J.P.; Spraul, M.; Neidig, P.; Nicholson, J.K. Chemometric Models for Toxicity Classification Based on NMR Spectra of Biofluids. *Chem. Res. Toxicol.* **2000**, *13*, 471–478. [CrossRef]
143. Lindon, J.C.; Nicholson, J.K.; Holmes, E.; Everett, J.R. Metabonomics: Metabolic processes studied by NMR spectroscopy of biofluids. *Concepts Magn. Reson.* **2000**, *12*, 289–320. [CrossRef]
144. Giraudeau, P.; Silvestre, V.; Akoka, S. Optimizing water suppression for quantitative NMR-based metabolomics: A tutorial review. *Metabolomics* **2015**, *11*, 1041–1055. [CrossRef]
145. Kostidis, S.; Addie, R.D.; Morreau, H.; Mayboroda, O.A.; Giera, M. Quantitative NMR analysis of intra- and extracellular metabolism of mammalian cells: A tutorial. *Anal. Chim. Acta* **2017**, *980*, 1–24. [CrossRef] [PubMed]
146. Wider, G.; Dreier, L. Measuring Protein Concentrations by NMR Spectroscopy. *J. Am. Chem. Soc.* **2006**, *128*, 2571–2576. [CrossRef] [PubMed]
147. Akoka, S.; Barantin, L.; Trierweiler, M. Concentration Measurement by Proton NMR Using the ERETIC Method. *Anal. Chem.* **1999**, *71*, 2554–2557. [CrossRef] [PubMed]
148. Bharti, S.K.; Roy, R. Quantitative 1H NMR spectroscopy. *TrAC Trends Anal. Chem.* **2012**, *35*, 5–26. [CrossRef]
149. Farrant, R.D.; Hollerton, J.C.; Lynn, S.M.; Provera, S.; Sidebottom, P.J.; Upton, R.J. NMR quantification using an artificial signal. *Magn. Reson. Chem.* **2010**, *48*, 753–762. [CrossRef]
150. Crockford, D.J.; Keun, H.C.; Smith, L.M.; Holmes, E.; Nicholson, J.K. Curve-Fitting Method for Direct Quantitation of Compounds in Complex Biological Mixtures Using 1H NMR: Application in Metabonomic Toxicology Studies. *Anal. Chem.* **2005**, *77*, 4556–4562. [CrossRef]
151. Singh, A.; Prakash, V.; Gupta, N.; Kumar, A.; Kant, R.; Kumar, D. Serum Metabolic Disturbances in Lung Cancer Investigated through an Elaborative NMR-Based Serum Metabolomics Approach. *ACS Omega* **2022**, *7*, 5510–5520. [CrossRef]

MDPI
St. Alban-Anlage 66
4052 Basel
Switzerland
Tel. +41 61 683 77 34
Fax +41 61 302 89 18
www.mdpi.com

Applied Sciences Editorial Office
E-mail: applsci@mdpi.com
www.mdpi.com/journal/applsci

www.ingramcontent.com/pod-product-compliance
Lightning Source LLC
LaVergne TN
LVHW070643170726
843515LV00004B/315